T0143361

Rapid Assessment Program

15

Bulletin of Biological Assessment

A Biological Assessment of the Aquatic Ecosystems of the Upper Río Orthon Basin, Pando, Bolivia

Barry Chernoff and Philip W. Willink, Editors

CONSERVATION INTERNATIONAL
THE FIELD MUSEUM
MUSEO NACIONAL DE HISTORIA
NATURAL - BOLIVIA

Bulletin of Biological Assessment is published by:
Conservation International
Center for Applied Biodiversity Science
Department of Conservation Biology
Rapid Assessment Program
2501 M Street NW, Suite 200
Washington, DC 20037
USA
202-429-5660
202-887-0193 fax
www.conservation.org

Editors: Barry Chernoff and Philip W. Willink
Assistant Editors: Leeanne E. Alonso and Jensen Reitz Montambault
Design: Glenda P. Fábregas
Maps: Philip W. Willink
Cover photograph: Theresa Bert
Translations: María del Carmen Arce S.

Conservation International is a private, non-profit organization exempt from federal income tax under section 501 c(3) of the Internal Revenue Code.

ISBN 1-881173-30-5
© 1999 by Conservation International.
All rights reserved.
Library of Congress Card Catalog Number 99-069140

The designations of geographical entities in this publication, and the presentation of the material, do not imply the expression of any opinion whatsoever on the part of Conservation International or its supporting organizations concerning the legal status of any country, territory, or area, or of its authorities, or concerning the delimitation of its frontiers or boundaries.

Any opinions expressed in the Bulletin of Biological Assessment are those of the writers and do not necessarily reflect those of Conservation International or its co-publishers.

Bulletin of Biological Assessment was formerly RAP Working Papers. Numbers 1-14 of this series were published under the previous title.

Suggested citation:
Chernoff, B. and P.W. Willink (Editors). 1999. A biological assessment of the aquatic ecosystems of the Upper Río Orthon basin, Pando, Bolivia. Bulletin of Biological Assessment 15. Conservation International, Washington, DC

PREFACE

From 4 - 20 September 1996 a team of 21 scientists undertook the first AquaRAP expedition to northern Bolivia, State of Pando, in the Tahuamanu and Manuripi river systems upstream from their confluence at the town of Puerto Rico (see Maps 1, 2 and 3). The scientific team was multinational and multidisciplinary. They came from Bolivia, Brazil, Perú, Paraguay, Venezuela and the United States. The scientists were specialists in aquatic and terrestrial botany; crustaceans and macro-invertebrates; population genetics; ichthyology; and limnology, including water chemistry and plankton. The AquaRAP team contained both established scientists and students. The expedition was focused primarily upon the biological and conservation value of the region as well as upon the current and future threats facing the region. However, the expedition was also intended to evaluate the protocols for rapidly assessing aquatic ecosystems.

The organization of this report begins with an executive summary which includes a brief overview of the physical and terrestrial characteristics of the region, as well as summaries of the chapters from each of the scientific disciplines, and concludes with recommendations for a conservation strategy. We then present the biological results from limnology, decapod crustaceans, fishes and genetics. The limnology group combines water chemistry with plankton and benthos. Although traditionally these disciplines have been separated, we combine what should be integrated. The disciplinary chapters are written as scientific papers, each self-contained with its own literature citations for ease of use. After the scientific papers we include a glossary of terms and the appendices with much of the raw data.

Explanations may be needed for the terms *diversity* and *species richness*. In vernacular usage *diversity*, when used in the context of biological organisms, refers to the numbers of and variety of types of species, organisms, taxa, etc. In the ecological literature, *diversity* takes on a slightly different meaning, referring to not only the number of entities but also their relative abundances. Such use of *diversity* can be found in Chapter 1. In this volume *diversity* is used in two ways: i) in the general, vernacular sense we occasionally refer to the "*diversity* of organisms", meaning the number of and variety of organisms; but ii) in the ecological sense, we refer to "low or high *diversity*", meaning "*diversity*" as calculated from a specific formula. Note that in the vernacular usage, *diversity* is never modified by high or low or used in a comparative way. With the exception of the limnological assessment, given the nature of rapid assessment we are unable to evaluate abundances of organisms in a quantitative fashion. Therefore, most of the disciplinary chapters reference only the number of species captured. From the ecological literature, the term *species richness* means the number of species. We use *species richness* in several chapters (e.g., Chapter 5) when referring to the number of species present in a habitat or river basin.

This report is intended to be used by decision makers, environmental managers, governmental and non-governmental agencies, students, and scientists. The novel information and analyses presented herein have two aims: (i) that we have presented a compelling case and cogent strategy for conservation efforts within the region; and (ii) that the scientific data and analyses will stimulate future scientific research of this critical region. We have attempted in this volume not simply to present an inventory of the organisms that we encountered during our expedition but rather to use that information to evaluate conservation strategies under different scenarios of environmental threat (e.g., Chapter 5). We welcome comments and criticisms as we continue to evolve AquaRAP and the methods used for evaluating conservation strategies from biological data.

Barry Chernoff
Philip W. Willink

TABLE OF CONTENTS

PARTICIPANTS

Francisco Barbosa (limnology)
Universidade Federal de Minas Gerais, ICB
Departamento de Biologia Geral
Lab. Limnologia/ Ecologia de Bentos
CP 486, CEP 30.161-970
Belo Horizonte, Minas Gerais, Brasil
Fax: 55-31-499-2567
Email: barbosa@mono.icb.ufmg.br

Soraya Barrera (ichthyology)
Museo Nacional de Historia Natural
Calle 26 s/n, Cota Cota
Casilla 8706
La Paz, Bolivia
Fax: 591-277-0876
Email: sbarrera@mail.megalink.com

Theresa Bert (genetics)
Florida Department of Environmental Protection
Florida Marine Research Institute
100 Eighth Ave., S.E.
St. Petersburg, FL 33701 USA
Fax: 813-823-0166
Email: bert_t@epic7.dep.state.fl.us

Barry Chernoff (ichthyology, AquaRAP Team Leader)
Department of Zoology
Field Museum
1400 South Lakeshore Dr.
Chicago, IL 60605-2496 USA
Fax: 312-665-7754
Email: chernoff@fmnh.org

Robin Foster (terrestrial botany)
Environmental & Conservation Programs
c/o Botany Department
Field Museum
1400 South Lakeshore Dr.
Chicago, IL 60605-2496 USA
Fax: 312-665-7932
Email: foster@fmnh.org

Antonio Machado-Allison (ichthyology)
Universidad Central de Venezuela
Instituto Zoologia Tropical
Laboratorio de Ictiologia
Apto Correos 47058
Caracas 10410-A, Venezuela
Fax: 058-2-605-2904
Email: amachado@strix.ciens.ucv.ve

Célio Magalhães (invertebrate zoology)
Instituto Nacional de Pesquisas da Amazonia
Coordenacao de Pesquisas em Biologia Aquatica
Caixa Postal 478
69.011-970 Manaus, AM, Brasil
Fax: 55-92-643-3095
Email: celiomag@inpa.gov.br

Naércio Menezes (ichthyology)
Museu de Zoologia
Universidade de São Paulo
Avenida Nazare 481
CEP: 04263-000
São Paulo, SP, Brasil
Fax: 55-011-274-3690
Email: naercio@usp.br

María Fatima Mereles (aquatic botany)
Departmento de Botánica
Facultad de Ciencias Químicas
Universidad Nacional de Asunción
P.O. Box PY 11001-3291
Campus UNA - Paraguay
Fax: 595-21-58-5564
Email: fmereles@sce.cnc.una.py

Debra Moskovits (coordinator)
Environmental and Conservation Programs
Field Museum
1400 South Lakeshore Dr.
Chicago, IL 60605-2496 USA
Fax: 312- 665-7932
Email: dmoskovits@fmnh.org

Hernán Ortega (ichthyology)
Museo de Historia Natural, UNMSM
Apartado 14,0434
Lima 14, Perú
Fax: 511-265-6819
Email: hortega@musm.edu.pe

Jaime Sarmiento (ichthyology)
Museo Nacional de Historia Natural
Calle 26 s/n, Cota Cota
Casilla 8706
La Paz, Bolivia
Fax: 591-277-0876
Email: mnhn@mail.megalink.com

ORGANIZATIONAL PROFILES

CONSERVATION INTERNATIONAL

Conservation International (CI) is an international, nonprofit organization based in Washington, DC. CI believes that the Earth's natural heritage must be maintained if future generations are to thrive spiritually, culturally, and economically. Our mission is to conserve the Earth's living heritage, our global biodiversity, and to demonstrate that human societies are able to live harmoniously with nature.

Conservation International
2501 M St., NW, Suite 200
Washington, DC 20037 USA
202-429-5660
202-887-0193 (fax)
http://www.conservation.org
http://www.conservation.org/AquaRAP

FIELD MUSEUM

The Field Museum (FM) is an educational institution concerned with the diversity and relationships in nature and among cultures. Combining the fields of Anthropology, Botany, Geology, Paleontology and Zoology, the Museum uses an interdisciplinary approach to increasing knowledge about the past, present and future of the physical earth, its plants, animals, people, and their cultures. In doing so, it seeks to uncover the extent and character of biological and cultural diversity, similarities and interdependencies so that we may better understand, respect, and celebrate nature and other people. Its collections, public learning programs, and research are inseparably linked to serve a diverse public of varied ages, backgrounds and knowledge.

Field Museum
1400 South Lakeshore Dr.
Chicago, IL 60605 USA
312-922-9410
312-665-7932 (fax)
http://www.fieldmuseum.org/

MUSEO NACIONAL DE HISTORIA NATURAL - BOLIVIA

The National Museum of Natural History (MNHN) is a scientific and educational institution that is funded by the National Academy of Sciences of Bolivia. The MNHN undertakes scientific investigations about the flora, fauna and paleontology of Bolivia in addition to its Department of Public Programs which maintains an exhibit hall. The Museum was created in 1980 and serves as an important center for scientific investigation.

In 1992, MNHN, in agreement with the Institute of Ecology of the Universidad Mayor de San Andrés, established the Bolivian Collecion of Fauna whose principal objective is to contribute to basic knowledge of the Bolivian fauna regarding biodiversity and distribution as well as information about conservation and sustainable use of the fauna. The Bolivian Collection of Fauna is the primary center for faunal collections and contains more than 125,000 specimens of vertebrates.

Museo Nacional de Historia Natural
Calle 26 s/n, Cota Cota; Casilla 8706
La Paz, Bolivia
591-279-5364
591-277-0876 (fax)
Email: mnhn@mail.megalink.com

ACKNOWLEDGMENTS

We would like to express our sincere gratitude to Kim Awbrey, formerly RAP Manager at CI, and to Lois "Lucho" James, Harmonia International. Without their hard work and foresight we would not have been able to accomplish what we did. Guillermo Rioja, formerly Director of CI-Bolivia, and much of his staff were instrumental in arranging permits and shipments, and helped the expedition in innumerable ways. In the field the motoristas and cooks, especially Doña Ana and Don Francisco, made accomodations and work as comfortable as possible. We would like to acknowledge the insights and valuable help from Adrian Forsyth, formerly Vice President of Conservation Biology of CI, and Debra Moskovits, Director of Environmental and Conservation Programs (ECP) at the Field Museum, both of whom were founding members of AquaRAP. Discussions with Adrian, Debby and Jorgen Thomsen, Vice President for Conservation Biology at CI, stimulated us to discover the approaches used in Chapter 5. Leeanne Alonso, RAP Director at CI, provided much needed assistance with scientific expertise and copy editing. Thanks also to the Rapid Assessment Program's Jensen Reitz Montambault and Debbie Gowensmith for their invaluable editorial assistance. Glenda Fábregas, CI, not only generated the page proofs for this volume but also extended to us more courtesies than we can ever thank her for. Funding for the expedition and development of AquaRAP protocols was generously provided by the W. Alton Jones Foundation. Expeditionary expenses were also contributed from the Marshall Field Funds, Department of Zoology, and ECP, Field Museum. We would also like to thank Jay Fahn, Mrs. Lee, David and Janet Shores and the 3-M Corporation for their generous contributions to AquaRAP. The Rufford Foundation has provided invaluable support for the further development on AquaRAP in South America. Lastly, we express our deepest gratitude to John McCarter, Russell Mittermeier, and Peter Seligmann for their continuing support of AquaRAP.

REPORT AT A GLANCE

A Biological Assessment of the Aquatic Ecosystems of the Upper Río Orthon Basin, Pando, Bolivia

1) Dates of Studies:
AquaRAP Expedition: September 4 - 20, 1996

2) Description of Location:
The region is a transition zone floristically between moister lowland amazonian forests to the north and east and the dryer deciduous forests to the south. The riparian forest communities and vegetation of the floodplains impart a unique character to the rivers and aquatic communities in this remote section of the upper Madeira river basin. The Ríos Tahuamanu and Manuripi join to form the Río Orthon which after a short distance anastomoses with the Madre de Dios and Beni rivers.

3) Reason for AquaRAP Studies:
This region in northern Pando, Bolivia is threatened by increasing human occupation and commercial activities in the area. Large tracts of forests within Pando are being converted to pastures for cattle, with the hardwood being removed for lumber. This habitat conversion puts great pressure on both terrestrial and aquatic ecosystems. Furthermore, there is an unregulated food fishery primarily for export to Brazil. For these reasons and because the Tahuamanu and Manuripi basins are close to the northern boundary of the Heath-Madidi conservation region, immediate attention was required.

4) Major Results:
The heterogeneity of the habitats, the uniqueness of the forests and the relative isolation of the Tahuamanu and Manuripi systems render this small region in northern Bolivia as potentially one of the richest in aquatic biodiversity within Bolivia if not within the Amazon River basin. This result was both unexpected and not obvious from scientific literature, maps or aerial photographs.

Zooplankton:	120 species
Benthic Macroinvertebrates:	19 orders and families
Chironomidae (Flies):	17 genera
Shrimps and Crabs:	10 species
Fishes:	313 species
Genetic Samples:	500 samples from over 50 fish genera

5) New Records for Bolivia:

Crabs:	3 species
Shrimps:	3 species
Fishes:	87 species

6) Conservation Recommendations:
It is critical to protect the Río Manuripi and Río Nareuda, which together represent >75% of the fish diversity within the region. Flooded areas must be protected as nurseries for fish and invertebrates. Igarapes (tributary streams) are most vulnerable and need protection.

INFORME DE UN VISTAZO

Una Valoración Biológica en la Cuenca Superíor del Río Orthon, Pando, Bolivia

1) Fechas de Estudio:
Expedición AquaRAP: Septiembre 4 al 20, 1996

2) Descripción del Sitio:
La región es una zona de transición florística entre los bosques húmedos de tierras bajas amazónicos hacia el norte y el este y los bosques secos deciduos hacia el sur. Las comunidades de bosque riparío y la vegetación de las planicies inundadas imparten una característica única a los ríos y comunidades acuáticas en este sector remoto de la cuenca superíor del Río Madeira. Los Ríos Tahuamanu y Manuripi se unen para formar el Río Orthon, que luego de una corta distancia confluye con los Ríos Madre de Dios y Beni.

3) Motivo de los Estudios AquaRAP:
En esta región en el norte de Pando, Bolivia está amenazada por el incremento de asentamientos humanos y actividades comerciales en el sector. Gran cantidad de bosque dentro de Pando está siendo convertida en pastizales para ganado, convirtiendo la madera dura cortada para la industria maderera. Esta conversión de hábitat pone mucha presión tanto en el sistema terrestre como acuático. Más aún, existe una pesca no regulada principalmente para su exportación al Brasil. Por estas razones y debido a que las cuencas del Tahuamanu y Manuripi se cierran hacia la frontera norte de la región de conservación Heath-Madidi, se requiere atención inmediata.

4) Resultados más Importantes:
La heterogeneidad del hábitat, la singularidad de los bosques y el aislamiento relativo de los sistemas Tahuamanu y Manuripi representan a esta pequeña región al norte de Boliva como una de las más ricas en biodiversidad acuática no solo dentro de Bolivia sino dentro de la cuenca Amazónica. Basándose en literatura, mapas o fotografías aéreas científicos, este resultado fue un tanto imprevisto.

Zooplancton:	120 especies
Macroinvertebrados Bénticos:	19 ordenes y familias
Chironomidae (Moscas):	17 géneros
Camarones y Cangrejos:	10 especies
Peces:	313 especies
Muestras Genéticas:	500 muestras de más de 50 géneros de peces

5) Nuevos Registros para Bolivia:

Cangrejos:	3 especies
Camarones:	3 especies
Peces:	87 especies

6) Recomendaciones de Conservación:
Es crítico proteger al Río Manuripi y al Río Nareuda, que en forma conjunta representan más del 75% de diversidad en peces dentro de la región. Las áreas inundadas deben ser protegidas como criaderos de peces. Los "Igarapes" (arroyos tributaríos) son lo más vulnerables y necesitan protección.

EXECUTIVE SUMMARY

INTRODUCTION

The area encompassing the Tahuamanu and Manuripi river systems in the northern state of Pando, Bolivia, has been a largely unexplored region in the upper Río Madeira river basin. The region is an important transition zone floristically between moister lowland amazonian forests to the north and east and the dryer deciduous forests to the south. The riparian forest communities and vegetation of the floodplains impart a unique character to the rivers and aquatic communities in this remote section of the upper Madeira river basin. The Tahuamanu and Manuripi Rivers join to form the Río Orthon[1], which after a short distance anastomoses with the Madre de Dios and Beni rivers (Map 1). The Beni and the Madre de Dios have much wider flood plains, and the Beni traverses incredible savannas and plains — ecotones not well represented in the Orthon system. Perhaps this fact as well as the heterogeneity of the habitats, the uniqueness of the forests, and the relative isolation of the Tahuamanu and Manuripi systems render this small region in northern Bolivia as potentially one of the richest in aquatic biodiversity within Bolivia if not within the Amazon River Basin. This result was both unexpected and not obvious from maps or aerial photographs.

Nonetheless, this region in northern Pando is coming under greater threat due to increasing human occupation and commercial activities in the area. Large tracts of forests within Pando are being converted to pastures for cattle, with the hardwood being removed for timber. This habitat conversion puts much pressure on both terrestrial and aquatic ecosystems. Furthermore, there is an unregulated food fishery primarily for export to Brazil. For these reasons and because the Tahuamanu and Manuripi basins form the northern boundary of the Heath-Madidi conservation region (Carr 1994), immediate attention was required.

[1] The name "Río Orthon" will be used in this report. This river has also been called the Orton and Ortho.

From 4 - 20 September 1996 a team of 21 scientists undertook the first AquaRAP expedition in northern Bolivia, State of Pando, in the Tahuamanu and Manuripi river systems upstream from their confluence at the town of Puerto Rico (Maps 2 and 3). The scientific team was multinational and multidisciplinary. They came from Bolivia, Brazil, Perú, Paraguay, Venezuela and the United States. The scientists were specialists in aquatic and terrestrial botany; crustaceans and macro-invertebrates; population genetics; ichthyology; and limnology, including water chemistry and plankton. The AquaRAP team contained both established scientists and students. Though the expedition was focused primarily upon the biological and conservation value of the region as well as upon the current and future threats facing the region, the expedition was also intended to evaluate the protocols for rapidly assessing aquatic ecosystems.

The South American Aquatic Rapid Assessment Program (AquaRAP) is a multinational, multidisciplinary program devoted to identifying conservation priorities and sustainable management opportunities in freshwater ecosystems in Latin America. AquaRAP's mission is to assess the biological and conservation value of tropical freshwater ecosystems through rapid inventories, and to report the information quickly to local policy makers, conservationists, scientists and international funding agencies. AquaRAP is a collaborative program managed by Conservation International and The Field Museum.

At the core of AquaRAP is an international steering committee composed of scientists and conservationists from seven countries (Bolivia, Brazil, Ecuador, Paraguay, Peru, Venezuela and the United States). The steering committee oversees the protocols for rapid assessment and the assignment of priority sites for rapid surveys. AquaRAP expeditions, which involve major collaboration with host country scientists, also promote international exchange of information and training opportunities. Information gathered in AquaRAP expeditions is released through Conservation International's Bulletin of Biological Assessment designed for local decision-makers, politicians, leaders, and conservationists, who can set conservation priorities and guide action through funding in the region.

We present here a brief overview of the physical and terrestrial characteristics of the region: the climate, the geology, the forest and the watershed. We then summarize the biological results from limnology, crustaceans, fishes and genetics. The limnology group combines water chemistry with plankton and benthos. Although traditionally these disciplines have been separated, we combine what should be integrated. Following the results, we recommend a conservation strategy for the region.

Climate

Mean annual precipitation in the Orthon basin is 1700 - 2000 mm/ year (Bartholomew et al., 1980; Killeen, 1998). Mean annual temperature is 25° to 26° C in eastern Bolivia, with October and November being the warmest months of the year, reaching 38° C, and July the coldest, reaching 10° C (Killeen, 1998). The wet season is October to April. The seasonality of the rains influences the water levels in the rivers, although there is a lag between the two because it takes time for all the water to enter into the rivers. Flooding in northern Bolivia occurs from December to May, with the peak reached in March (Goulding, 1981).

The primary climatological phenomenon for the region is interaction of southern cold air masses (surazos) and the intertropical convergence zone (ITCV) (Killeen, 1998). During the wet season, the ITCV is responsible for warm, moist air passing westward over the Amazon basin until it nears the Andes, at which point the air masses move southward. It is this weather pattern that is responsible for the seasonal rains and subsequent flooding. During the dry season, cold, dry air from the south forces the ITCV northward. This results in reduced rainfall (Killeen, 1998). Although the surazos are most common in the winter, they can appear year-round, giving rise to sudden and fierce storms. The interaction between the surazos and ITCV is influenced by El Niño (Killeen, 1998). It is unknown exactly how this affects the weather of northern Bolivia, but it is possible that rainfall is above average during El Niño years and below average during La Niña years due to shifting of the ITCV (Killeen, 1998).

Geology

The Upper Río Orthon is part of the Río Madeira basin. The Río Madeira basin is located between the Andes and the Brazilian Shield and accounts for one-fifth of the area of the Amazon basin. Regional rivers are believed to have flowed northwest during the early Tertiary (Goulding, 1981). With the rise of the Andes in the Miocene, the rivers probably emptied into an inland sea, the Pebasian Sea, which was connected to the Caribbean (Räsänen et al., 1995; Hoorn, 1996; Marshall and Lundberg, 1996). The present day Amazon formed sometime in the late Miocene, when the Madeira began to take on its present form (Hoorn 1996). Currently, the Madeira flows through a pass across the Brazilian Shield. This pass is relatively narrow, forcing Bolivian water to back up during the wet season, enhancing the flooding in Bolivia (Goulding, 1981).

The headwaters of the Madre de Dios are separated from other Amazon headwaters by the Fitzcarrald arch, a ridge where the crustal plates have buckled along the Río Purus due to Andean development (Räsänen et al., 1987). From the Permian to the Quaternary, the buckling formed depressions, including one at the headwaters of the Madre de Dios. This depression was filled in over time by riverine deposits (Räsänen et al., 1987).

The Brazilian Shield is granitic Precambrian rock (Killeen, 1998), but most of the region is covered by soil eroded from the Andes during the Tertiary (Ahlfeld, 1972; Bartholomew et al., 1980). The hard rock of the shield is exposed in some areas and forms the basis for the regional rapids. Unlike the lower Río Madeira in the vicinity of the Amazon, the river floodplains in northern Bolivia are very narrow (Goulding, 1981). The current Upper Río Orthon is of Andean origin, hence the waters are often laden with sediments and moderate in nutrients (whitewater). Blackwaters are also present, as noted in some of the smaller tributaries passing through terra firme forests (e.g. Garape Preto).

CHAPTER SUMMARIES

Water Quality, Zooplankton, and Benthic Macroinvertebrates

Water quality was assessed at 22 sites. Waters were generally white, slightly acidic to neutral, and well oxygenated. The concentration of nutrients was medium, and temperatures ranged between 19° and 31° C. There were also some blackwater habitats, as well as localities where oxygen was depleted (e.g. shallow lakes). In general, water was of good quality and did not exhibit significant signs of contamination or eutrophication.

Zooplankton communities were assessed at 10 localities. A total of 120 taxa were identified. Rotifera had the highest species richness (44%), although Protozoa (40%) were also diverse. Cladocera, Copepoda, Gastrotricha, Ostracoda, and Nematoda were also represented. Taxa could be classified as planktonic, semi-planktonic, or littoral/ benthic. Most have cosmopolitan or tropical distributions.

Benthic macroinvertebrate communities were assessed at 8 localities. A total of 1,665 organisms were identified. Bivalvia were most numerous (27%), followed by Chironomidae, Heteroptera, and Oligochaeta. Gastropoda, Hirudinea, Decapoda, Aracnida, Odonata, Coleoptera, and Diptera were also represented. Bivalvia were most common in neutral to slightly alkaline waters. Among Chironomidae, 17 taxa were identified, and most of these are associated with the sediments. The benthic macroinvertebrate fauna showed a high density of predators, although herbivores were also collected. Based on Shannon diversity indices and work done in the upper/middle Río Paraguay, six sites were considered to have intermediate diversity and two sites had low diversity. It is hypothesized that no sites were categorized as high diversity because small taxa were not collected due to limitations in sampling methods.

Crustaceans

The decapod crustacean fauna was assessed based on material collected by the two fish teams from 45 stations. Ten species of shrimps and crabs, representing three families and six genera, were found. Of these, six are new records for Bolivia. All the species are widely distributed throughout the Amazon basin and the neotropics, and are typically found in white water conditions. Several species previously recorded from Bolivia were not found, and this may be due to inadequate sampling methods. There were only minor differences in decapod community structure among the sub-basins. There were specific microhabitat associations, with some species (e.g. *Macrobrachium amazonicum*) preferring aquatic vegetation, others (e.g. *Macrobrachium brasiliense*) preferring submerged litter and dead trunks, and still others (e.g. *Macrobrachium jelskii*) with no readily discernable preferences. Shrimps and crabs play many important roles in aquatic and terrestrial food chains. It is anticipated that they can be conserved as long as their habitats are not significantly modified.

Fishes

Fishes were sampled at 85 stations and 313 species were collected, of which 87 are new records for Bolivia. This brings the total fish fauna of Bolivia to 641 species and for the Bolivian Amazon to 501 species. This small region in northeastern Bolivia contains 63% and 49% of all the species known to inhabit the Bolivian Amazon and Bolivia, respectively. This region is potentially a hotspot of fish biodiversity.

The distribution of fishes in the Upper Río Orthon basin has important ramifications for regional conservation recommendations. Simpson's index of similarity and a measure of matrix disorder (or entropy) were used to test preliminary hypotheses about the homogeneity of fish distributions among geographic subregions, macrohabitats, or class of water. Species were distributed non-randomly in regards to subregion, with highest number of species located in the Río Manuripi. It is believed to be a hierarchical pattern with the faunal similarities nested within the larger fauna represented by the Río Manuripi. Species were also distributed non-randomly in relation to macrohabitat; the largest numbers of species were found in riverine habitats. We could not determine unambiguously if this pattern was clinal or nested because there was no obvious way to order relationships among habitats. The data are most consistent with fauna being derived from riverine habitats on seasonal flooding cycles. Species were distributed homogeneously among black, white, and turbid waters. This finding was unexpected. Based on the distribution of fishes, it is recommended that the Río Manuripi and Río Nareuda be designated as core conservation areas because, combined, they represent 75% of the regional diversity. Tributaries are the most endangered macrohabitat because of their fragility, uniqueness, and the current trends in habitat destruction. More than 80% of the fishes are dependent upon the flooded areas for reproduction, nursery grounds, or critical food getting, which highlights the importance of seasonal inundations on the regional aquatic ecology.

A large percentage of the fauna have high economic value as food or ornamental fishes. Stocks are in need of immediate evaluation, and the exportation of food fishes needs to be regulated as soon as possible.

Genetics

Five hundred tissue or whole-body samples were collected from approximately 50 genera of fishes from the Manuripi River drainage for subsequent analysis of population genetic structure or phylogenetic relationships. Samples were frozen in liquid nitrogen or preserved in 100% ethanol. Voucher specimens were preserved in formalin. For most putative species, one to a few individuals were sampled; for a few species, multiple collections suitable for population genetic analysis were taken. We are in the process of genetically analyzing these samples. The results of that work will be available in the future. The structure of the Manuripi River drainage system is heterogeneous and should provide ample opportunity for natural selection to promote processes of differentiation. The variation in presence and relative abundance of common species found in different streams and cochas provides an opportunity for natural selection to act on some fish species, which could contribute to the maintenance of a relatively high level of genetic diversity within these species. From a genetic perspective, perhaps the most important river basin to preserve is the Río Manuripi. Our preliminary evaluation suggests that it is important to preserve both the water quality and the physiography of this basin. This, of course, necessitates the preservation, at least to some degree, of the entire drainage system.

RECOMMENDED CONSERVATION AND RESEARCH ACTIVITIES

The upper Río Orthon basin of Bolivia – also referred to as the Tahuamanu-Manuripi region – was discovered to be a potential hotspot of aquatic biodiversity and, therefore, is a high priority for immediate study and conservation. The basis for the following recommended conservation and research activities can be found in the text of the executive summary and the various chapters. Recommendations are not listed in order of priority.

- More faunal and floristic surveys are required. Even at the conclusion of this AquaRAP expedition, additional species were being collected. Our knowledge of the regional biota is still incomplete.

- It is critical to maintain the hydrological cycle responsible for the annual flooding which creates and maintains the lagoons and backwaters. These lagoons and backwaters serve as nursery and feeding areas for a large number of invertebrates and fishes, provide sheltered areas for the growth of aquatic plants, and likely serve as sites that may contribute to genetic diversity. Rising water levels cue many species of fishes to begin reproducing. Dams and channelization have a disastrous impact on their riverine communities.

- Zones of critical habitats with varzea, cochas, main channels and upland areas need to be created and well protected. Rather than the formation of a park, multiple use zones with some habitats restricted for modification should be considered. Special attention needs to be focused on the flood zone. The narrow nature of the flood zone brings human activities closer to, and is now infringing upon, the varzeas, cochas and flooded areas. As mentioned elsewhere, seasonal flooding is a critical phenomenon for the local aquatic ecosystem.

- It is critical to designate two areas within the Tahuamanu-Manuripi region as core conservation areas: the Río Manuripi and the Río Nareuda. The majority of the regional biodiversity can be found in these two areas. Any degradation to the Manuripi system will require setting a number of alternate zones, such as the Middle Río Tahuamanu and the Lower Río Nareuda as new core zones.

- Populations and stocks of commercially exploited fishes need to be studied immediately across the Río Madeira basin in Bolivia and coordinated with Brazil. Although some species serve as a food source for local people, much of the fishery may largely be for exportation. This exploitation of local fishes could be putting undo pressure on stocks, for catch data from the expedition do not indicate that viable populations of key fish species exist in the region. Management of the fishery may be required to prevent a collapse.

- Restoration of the gallery forest and restriction of burning is highly recommended. The Upper and Middle Tahuamanu has been severely damaged by burning and clearing land for cattle ranching. Ashes may poison the waters. The small size of tributary streams, such as the igarapes, and their high dependence upon riparian vegetation make them extremely susceptible to the clearing of land. Cattle ranching is likely to increase, not decrease, throughout the entire region.

- We recommend working with local fisherman and floodplain residents to explain the relationships between maintenance of habitats and biodiversity of fishes. Develop educational programs and involve local residents and fishermen in monitoring stocks and habitats. Educational programs should promote the learning of the different fishes and how to recognize new or unusual species that should be brought to the attention of scientists.

- More basic ecological studies are needed to gain a better understanding of the interdependencies of the region's flora and fauna. For example, shrimps and crabs are a critical component of the food chain, both as predators and prey. This food chain is not restricted to the aquatic ecosystem since many birds, reptiles, and mammals also feed on aquatic organisms.

- The potential to develop local fishery for ornamental species should be studied. Many fish species are of high value in the aquarium trade and could be a source of income for the local people. This activity would best be conducted in the isolated lagoons, cochas, or dead arms of the river, but population biology and life histories need to be studied to guarantee sustainability. Managed properly, this would also help to promote the conservation of the aquatic ecosystem.

- Rivers provide water for the local inhabitants. Although not a widespread problem, pollution and waste treatment near larger population centers needs to be regulated to prevent contamination of drinking water as populations grow.

LITERATURE CITED

Ahlfeld, F.E. 1972. Geología de Bolivia. Los Amigos del Libro, La Paz, Bolivia. 190 pp.

Bartholomew, J.C., P.J.M. Geelan, H.A.G. Lewis OBE, P. Middleton, and B.Winkleman. 1980. The times atlas of the world: comprehensive edition. Time Books, London. 227 pp.

Foster, R.B., J.L.Carr, and A.B. Forsyth (eds.) The Tambopata-Candamo Reserved Zone of southeastern Perú: a biological assessment. RAP Working Papers No. 6, Conservation International, Washington, D.C. 184pp.

Goulding, M. 1981. Man and fisheries on an Amazon frontier. Dr. W. Junk Publishers, Boston. 137 pp.

Hoorn, C. 1996. Miocene deposits in the Amazonian foreland basin. Science 273:122.

Killeen, T.J. 1998. Introduction. *In* Killeen, T.J., and T.S. Schulenberg (eds.). A biological assessment of Parque Nacional Noel Kempff Mercado, Bolivia. Pp. 43-51. RAP Working Papers No. 10 Conservation International, Washington D.C. 372 pp.

Marshall, L.G., and J.G. Lundberg. 1996. Miocene deposits in the Amazonian foreland basin. Science 273: 123-124.

Räsänen, M.E., A.M. Linna, J.C.R. Santos, and F.R. Negri. 1995. Late Miocene tidal deposits in the Amazonian foreland basin. Science 269: 386-390.

Räsänen, M.E., J.S. Salo, and R.J. Kalliola. 1987. Fluvial perturbance in the western Amazon basin: regulation by long-term sub-Andean tectonics. Science 238: 1398-1401.

RESUMEN EJECUTIVO

INTRODUCCION

El área que comprende los sistemas de los ríos Tahuamanu y Manuripi al norte del Departamento de Pando, Bolivia, ha sido una gran región no explorada en la cuenca superíor del Río Madeira. La región es una zona importante de transición florística entre los bosques húmedos de tierras bajas amazónicas hacia el norte y el este y bosques secos deciduos hacia el sur. Las comunidades de bosque riparío y la vegetación de las planicies inundadas imparten un caráctaer único a los ríos y comunidades acuáticas en este sector remoto de la cuenca superíor del Río Madeira. Los Ríos Tahuamanu y Manuripi se unen para formar el Río Orthon, que luego de una distancia corta confluye con los ríos Madre de Dios y Beni (Mapa 1). El Río Beni y el Río Madre de Dios tienen planicies aluviales mucho más amplias y el Río Beni atravieza increibles sabanas y planicies — ecotonos no bien representados en el sistema Orthon. Tal vez este detalle como la heterogeneidad de los hábitats, la singularidad de los bosques, y el aislamiento relativo de los sistemas Tahuamanu y Manuripi representan para esta pequeña región del norte de Bolivia como una de las más ricas en biodiversidad acuática no solo en Bolivia si no en toda la cuenca del Río Amazóna. Este análisis es poco evidente de acuerdo a los mapas y fotografías aéreas.

Sin embargo, esta región del norte de Pando está bajo mucha amenaza, debido al crecimiento de asentamientos humanos y actividades comerciales en el área. Grandes franjas de bosque dentro de Pando se están convirtiendo en pastizales para ganado, con la explotación de árboles para la industria maderera. Esta conversión del hábitat pone mucha presión tanto en el ecosistema terrestre como acuático. Más aún, existe una pesca alimenticia no regulada principalmente para su exportación al Brasil. Por estas razones y debido a que las cuencas del Tahuamna y el Manuripi forman la frontera norte de la region de conservación de Heath-Madidi (Carr 1994), se requiere atención inmediata.

Del 4 al 20 de septiembre de 1996, un equipo de 21 científicos se hicieron cargo de la expedicion AquaRAP en el norte de Bolivia, Departamento de Pando, en los sistemas de los Ríos Tahuamanu y Manuripi arriba de su confluencia en el pueblo de Puerto Rico (Mapas 2 y 3). El equipo científico era multinacional y multidisciplinarío. Participaron equipos de Bolivia, Brasil, Perú, Paraguay, Venezuela y U.S.A. Los científicos eran especialistas en botánica terrestre, crustaceas y macro-invertebrados, genética de poblaciones, ictiología y limnología, incluyendo química del agua y plancton. El equipo AquaRAP fue conformaba tanto por científicos como estudiantes. Si bien la expedición se enfocó especialmente en el valor biológico y de conservación de la región, se contemplaron también las amenazas actuales y futuras que se enfrentarían en esta región. La expedición estaba dirigida a evaluar la metodología y protocolos, para evaluación rápida de ecosistemas acuáticos.

El Programa de Evaluación Rápida de Ecosistemas Acuáticos de Sudamérica (Aqua RAP) es un programa internacional y multidisciplinarío. El objetivo es el de determinar puntos príoritaríos de conservación y de mantenimiento sostenible en ecosistemas de agua dulce a lo largo de América Latina. Su función es la de evaluar el valor biológico y la necesidad de conservación de ecosistemas acuáticos tropicales por medio de inventaríos rápidos y hacer llegar los resultados sin demora a los responsables de politicos, científicos, conservacionistas y agencias internacionales. AquaRAP es un programa de colaboración entre Conservación Internacional y The Field Museum.

La estructura de AquaRAP es un comité formado por un equipo internacional de cientificos de siete paises (Bolivia, Brasil, Ecuador, Paraguay, Perú, Venezuela, y los

[1] El nombre "Río Orthon" se usará es este informe. Este río también lo han llamado Orton y Ortho.

Estados Unidos). El comité revisa y controla el protocolos y parametros utilizados para la selección de los puntos prioritarios de conservación para evaluaciones rápidas. Las expediciones, las cuales implican mayor colaboración entre los científicos de América Latina, también promueve un intercambio internacional y oportunidades de capacitación. La información recogida en las expediciones de AquaRAP es repartida por medio de una seria de Publicaciones de Evaluación Rápida de Conservación Internacional que fue diseñada para los responsables de decisiones locales, los politicos, los lideres y los conservacionistas, quienes pueden imponer prioridades de conservación y guiar acción es através de los fondos que llegan a la región.

Ahora presentamos una breve visión de las características físicas y terrestres de la región: el clima, la geología, el bosque y las cuencas. Luego resumiremos los resultados biológicos de limnología, crustáceos, peces y genética. El grupo de limnología combina la química del agua con plancton y bentos. Aunque tradicionalmente estas disciplinas han sido separadas, combinamos lo que debería ser integrado. Posteriormente, recomendamos una estrategia de conservación para la región.

Clima

La precipitación anual principal en la cuenca Orthon es de 1700 - 2000 mm/año (Bartholomew et al., 1980). La temperatura anual prinicipal es de 25° a 26°C al este de Bolivia, siendo Octubre y Noviembre los meses más cálidos del año, alcanzando 38°C, y julio el más frío, alcanzando 10°C (Killeen, 1998). La estación húmeda es de octubre a abril. La estacionalidad de las lluvias influye sobre los niveles de agua en los ríos. Si bien existe un intervalo entre las dos estaciones, toma tiempo para que todo el agua fluya en los ríos. La inundación en el norte de Bolivia ocurre de diciembre a mayo, con un pico que se alcanza en marzo (Goulding, 1981).

El fenómeno climatológico primarío para la región es la interacción de las masas de aire frío del sur (surazos) y la zona de convergencia intertropical (ITCV) (Killeen, 1998). Durante la estación húmeda, el ITCV es responsable del aire tibio húmedo que pasa en sentido oeste sobre la cuenca amazonica hasta que se acerca a los Andes, punto en el que las masas de aire se mueven hacia el sur. Este patrón climático es el responsable de las lluvias estacionales e inundaciones subsecuentes. Durante la estación seca, el aire frío y seco del sur fuerzan al ITCV ir hacia el norte. Esto resulta en lluvia reducida (Killeen, 1998). Aunque los surazos son más comunes en el invierno, ellos pueden aparecer a lo largo del año, causando tormentas repentinas y fuertes. La interacción entre los surazos y el ITCV es influenciada por El Niño (Killeen, 1998). Se desconoce exactamente como esto afecta el clima en el norte de Bolivia, pero es posible que la lluvia esté sobre el promedio durante los años de El Niño y bajo promedio durante los años de La Niña debido al cambio continuo del ITCV (Killeen, 1998).

Geología

La parte superíor del Río Orthon es parte de la cuenca del Río Madeira. La cuenca del Río Madeira está ubicado entre los Andes y el escudo del Brasil y se atribuye un quinto del área de la cuenca amazónica. Se cree que los ríos regionales fluyeron hacia el noroeste durante los comienzos del terciarío (Goulding, 1981). Con el levantamiento de los Andes en el Mioceno, los ríos probablemente se vaciaron a un mar interno, el Mar Pebasio, que estaba conectado al Caribe (Räsänen et al., 1995; Hoorn, 1996; Marshall y Lundberg, 1996). La actual Amazonia fue formada en algun momento durante el Mioceno, cuando el Río Madeira empezó a adquirir su forma actual (Hoorn, 1996). Actualmente, el Río Madeira fluye a través de un estrecho cañadon, que cruza el escudo brasileño. Este cañadon es relativamente angosto, dando lugar a que las aguas del lado boliviano retrocedan durante la estación húmeda, incrementando las inundaciones en Bolivia (Goulding, 1981).

Las aguas principales del Madre de Dios están separadas de otras aguas principales amazónicas por el arco Fitzcarrald, una serranía disectada, donde las placas de corteza terrestre han sido sujetadas a lo largo del Río Purus debido al desarrollo Andino (Räsänen et al., 1987). Desde el Pérmico al Cuaternarío, la sujeción formaba depresiones, incluyendo uno en las aguas principales del Madre de Dios. Esta depresión fue llenada a lo largo del tiempo por el depósito ribereño. (Räsänen et al., 1987).

El Escudo Brasilero está conformado por piedra granítica Precámbrica (Killeen, 1998) pero, la mayoría de la región está cubierta por suelo erosionado de los Andes durante el Terciarío (Ahlfeld, 1972; Bartholomew et al., 1980). La roca madre del Escudo está expuesta en algunas áreas, y forma la base de los "rápidos" regionales. A diferencia de aguas abajo del Río Madeira en la vecindad de la Amazonía, las planicies inundadas de río en el norte de Bolivia son muy angostas (Goulding, 1981). La corriente río arriba del Orthon es de origen andino, por lo tanto las aguas son frecuentemente llenas de sedimentos y moderadas en nutrientes (aguas blancas). También se advierten aguas negras, como se observa en tributaríos más pequeños que pasan a través de bosque de tierra firme (e.g. Garape Preto).

RESUMEN DE CAPITULOS

Calidad de Agua, Zooplancton, y Macroinvertebrados Benticos

La calidad de agua fue registrada en 22 sitios. Las aguas generalmente eran blancas, ligeramente ácidas a neutrales, y bien oxigenadas. La concentración de nutrientes era media y la temperatura estaba en un rango entre 19° a 31°C. También habían algunos hábitats de aguas negras, como también localidades donde el oxígeno era reducido o agotado (e.g. lagos superficiales). En general, el agua era de buena calidad y no exibía señales significantes de contaminación o eutrofización.

Las comunidades de Zooplancton fueron registradas en 10 localidades. Un total de 120 taxa fueron indentificados. La Rotifera tenía la mayor riqueza en especies (44%), aunque Protozoa (40%) también era diversa. Cladocera, Copépoda, Gastrotricha, Ostracoda y Nematoda también estaban representadas. Los taxa podrían ser clasificados como planctónica, semi-planctónica, o litoral/béntica. La mayoría tienen distribuciones cosmopolitas o tropicales.

Las comunidades de macroinvertebrados bénticos fueron registrados en 8 localidades. Un total de 1,665 organismos fueron identificados. Bivalvia fue la más numerosa (27%) seguida de Chironomidae, Heteróptera y Ologochaeta. Gastrópoda, Hirudinea, Decápoda, Arácnida, Odonata, Coleóptera y Díptera bien representados. Bivalvia era lo más comun en aguas ligeramente alcalinas a neutrales. Dentro de las Chironomidae, 17 taxa fueron identificadas y la mayoría de estas están asociadas con los sedimentos. La fauna de macroinvertebrados benticos mostraron una alta densidad de predadores, aunque también se colectaron herbívoros. Basados en los Indices de Diversidad Shannon y el trabajo realizado en los río arriba/medio del Río Paraguay, seis sitios fueron considerados por contener diversidad intermedia y dos sitios por baja diversidad. Es una hipótesis que ningún sitio fue categorizado como de alta diversidad debido a que los taxa pequeños no fueron colectados, por las limitaciones en los métodos de muestreo.

Crustáceos

La fauna de crustáceos decápodos fue representada en base a material recogido por dos equipos de pesca en 45 estaciones. Diez especies de camarones y cangrejos, representantes de tres familias y seis géneros fueron encontradas. De estas, seis son registros nuevos para Bolivia. Todas las especies son ampliamente distruibuidas a lo largo de la cuenca amazónica y los neotrópicos, y son típicamente característicos de ambientes de aguas blancas. Varias especies anteriormente registradas en Bolivia no fueron encontradas, y esto se puede deber a métodos inadecuados de muestreo. Hubieron diferencias menores en la estructura de la comunidad de decápodos entre las subcuencas. Habían asociaciones especificas de microhábitat, con algunas especies (e.g. *Macrobrachium amazonicum*) que prefieren vegetación acuática, otros (e.g. *Macrobrachium brasiliense*) prefiriendo material sumergido y troncos muertos, y aún otros (e.g. *Macrobrachium jelskii*) con preferencias no disernibles. Los camarones y cangrejos tienen varios roles importantes en cadenas alimenticias acuáticas y terrestres. Se deduce que ellos pueden ser conservados si sus habitats no son modificados significativamente.

Peces

Los peces fueron muestreados en 85 estaciones y 313 especies fueron colectadas, de las cuales 87 son nuevos registros para Bolivia. Esto nos brinda un total de 641 especies de fauna de peces para Bolivia y 501 especies para la Amazonia Boliviana. Esta pequeña región en el noreste de Bolivia contiene el 63% y el 49% de todas las especies conocidas que habitan la Amazonia Boliviana y Bolivia respectivamente. Esta región es potencialmente un punto importante para la biodiversidad de peces.

La distribución de peces río arriba de la cuenca del Orthon tiene ramificaciones importantes para las recomendaciones regionales de conservación. El Indice de Simpson de similaridad y una medida de matriz desordenada (o entropia) fueron utilizados para examinar las hipótesis preliminares sobre la homogeneidad de distribución de peces en subregiones geográficas, macrohábitats, o clase de agua. Las especies fueron distribuidas uniformemente con respecto a la subregion, con un número más alto de especies ubicadas en el Río Manuripi. Se piensa que existe un patrón jerárquico con las similitudes de fauna encontrados dentro de la fauna mayor representada por el Río Manuripi. Las especies también fueron distribuidas uniformemente con relación al macrohábitat; la gran cantidad de especies fue encontrada en los hábitats ribereños. No pudimos determinar, si este patrón era clinal o anidado ya que no habia una manera obvia que ordene las relaciones entre los hábitats. Esta información es más consistente con fauna que deriva de hábitats ribereños en ciclos de inundación estacionales. Las especies fueron distribuidas en forma homogénea en aguas negras, blancas y turbias. Este descubrimiento fue inesperado. Basados en la distribución de peces, se recomienda que el Río Manuripi y el Río Nareuda sean asignados como áreas de conservación centrales ya que, combinados, representan el 75% de la diversidad regional. Los afluentes son los macrohábitats más amenazados debido a su fragilidad, singularidad, y las tendencias actuales de destrucción del hábitat. Más del 80% de los peces dependen de las áreas inundables para su reproducción, áreas semilleras, o de obtención crítica de alimento, lo que destaca la importancia de las inundaciones estacionales en la ecologia acuática regional. Un gran porcentaje de la fauna tiene un alto valor economico como alimento o comercialmente como peces ornamentales. Es necesaria una evaluación inmediata de reservas y regulaciones de exportación de pescado como alimento.

Genética

Quinientas muestras de tejido o especimen fueron colectadas de aproximadamente 50 géneros de peces del drenaje del Río Manuripi para el respectivo análisis de su estructura genética de población o relaciones phylogeneticas. Se congelaron algunas muestras en nitrógeno líquido o fueron preservados en 100% de ethanol. Las especies ejemplares fueron preservados en formol. Para la mayoría de las especies putativas, pocos individuos fueron muestreados; para algunas especies, se tomaron múltiples colectas adecuadas para el análisis de genética poblacional. Estamos en el proceso de analizar genéticamente estas muestras. Los resultados de este trabajo estarán disponible en el futuro. La estructura del sistema de drenaje del Río Manuripi es heterogéneo y debería proveer una amplia oportunidad para la selección natural a manera de promover procesos de diferenciación . La variación en la presencia y abundancia relativa de especies comunes encontradas en diferentes arroyos y cochas, suministran una oportunidad para la selección natural de actuar sobre algunas especies de peces que podría contribuir al mantenimiento de un nivel alto de diversidad genética dentro de estas especies. Desde un punto de vista genético, tal véz la cuenca de río más importante para preservar, es el Río Manuripi. Nuestra evaluación preliminar sugiere que es importante preservar tanto la calidad de agua como la fisiografía de esta cuenca. Esto, por supuesto, necesita al menos cierto grado de preservación del completo sistema de drenaje.

ACTIVIDADES RECOMENDADAS PARA LA CONSERVACION E INVESTIGACION

La cuenca río arriba del Orthon de Bolivia – también referido como la región Tahuamanu-Manuripi – fue descubierta como un punto potencial de biodiversidad acuática y, por lo tanto, de alta príoridad para el estudio y conservación inmediato. La base para las siguientes actividades recomendadas de conservación e investigación se encuentran en el texto del resumen ejecutivo de los capitulos. Las recomendaciones no están enumeradas en orden de príoridad.

- Se requieren más relevamientos de campo en fauna y flora. Aún sobre la conclusión de esta expedición de AquaRAP, especies adicionales fueron colectadas. Nuestro conocimiento de la biota regional aún está incompleta.

- Es crítico mantener el ciclo hidrológico, debido al cual se produce la inundación anual que origina y mantiene las lagunas y las aguas negras. Estas lagunas y aguas negras sirven como un criadero y área de alimentación para un gran número de invertebrados y peces, proveen refugio para el crecimiento de plantas acuáticas, y sirven como sitios que pueden mantener la diversidad genética. Al subir el nivel del agua se da lugar a la migración de muchas especies de peces para dar inicio al proceso de reproducción. Los diques y canalizaciones tienen un impacto desastroso en las comunidades ribereñas.

- Zonas de hábitats críticos con varzea, cochas, canales principales y tierra firme necesitan estar bien protegidos. Más que el establecimiento de un Parque, se debe considerar zonas de uso múltiple con algunos hábitats restringidos para posteríor modificación. Se necesita prestar atención especial en la zona inundada. Por la problemática que conlleva la actividad humana más cercana a la zona inundada, las cuales estan siendo afectadas, como las varzeas, cochas y áreas inundadas, las inundaciones estacionales son un fenómeno crítico para el ecosistema acuático local, como se indicó anteriormente.

- Es crítico designar dos áreas dentro de la región Tahuamanu-Manuripi como áreas centrales de conservación: el Río Manuripi y el Río Nareuda. La mayoría de la biodiversidad regional puede ser encontrada en estas dos áreas. Cualquier degradación del sistema Manuripi requerirá el establecimiento de un número de zonas alternativas, como ser el centro del Río Tahuamanu y aguas abajo del Río Nareuda como zonas centrales nuevas.

- Las poblaciones y reservas de peces comercialmente explotados, necesitan ser estudiados inmediatamente a través de la cuenca del Río Madeira en Bolivia, y coordinados con Brasil. Aunque algunas especies sirven como un recurso alimenticio para la gente del lugar, la mayor parte de la pesca puede ser para la exportación. Esta explotación de peces locales pueden producir una presión innecesaria a las reservas, ya que los datos de la expedición no indican que existan poblaciones viables de las especies principales de peces en la región. La administración de pesca puede ser requerida para prevenir un colapso.

- La restauración del bosque de galería y la restricción de quema son altamente recomendables. El Tahuamanu alto y central han sido severamente dañados por las quemas y apertura de tierras para la ganadería. La ceniza puede envenenar las aguas. El tamaño pequeño de los arroyos tributarios, como ser los "igarapes", y su alta dependencia de la vegetacion riparina las convierte en extremedamente susceptibles a la tala de árboles. La ganadería continua incrementándose, a través de toda la región.

- Recomendamos trabajar con las comunidades locales de pescadores, especialmente los cercanos a las planicies inundables, para explicar las relaciones entre el mantenimiento de los hábitats y la biodiversidad de los peces. Se deben desarrollar programas educativos e involucrar a los habitantes en monitorear las reservas de peces y sus hábitats. Los programas educativos deben promover el aprendizaje sobre los diferentes peces y como reconocer especies nuevas o raras que debería llamar la atención de los científicos.

- Mayores estudios ecológicos básicos son necesaríos para obtener un mejor entendimiento de las interdependencias de la flora y fauna de la región. Por ejemplo, camarones y cangrejos son un componente crítico de la cadena alimenticia, dado que ambos cumplen funciones de predadores y presa. Esta cadena alimenticia no está restringida al ecosistema acuático, ya que muchas aves, reptiles, y mamíferos también se alimentan de organismos acuáticos.

- Estudiar el potencial para desarrollar una pesca local para peces ornamentales. Muchas especies de peces tienen un alto valor en el intercambio de acuarios, y pueden ser un recurso de ingreso para los habitantes locales. Esta actividad sería mejor conducida en las lagunas aisladas, cochas, o ramificaciones del río. Tambien, la biología de población y la historia natural necesitan ser estudiadas para garantizar la sostenibilidad. Con una buena administración, esto también ayudaría a promover la conservación del ecosistema acuático.

- Los ríos suministran agua para los habitantes locales. Aunque no es un problema muy evidente todavía, el tratamiento de la basura cerca de los centros poblados en crecimiento necesita ser regulado para prevenir la contaminación de agua potable a la población.

LITERATURA CITADA

Ahfeld, F.E. 1972. Geología de Bolivia. Los Amigos del Libro, La Paz, Bolivia. 190 pp.

Bartholomew, J.C., P.J.M. Geelan, H.A.G. Lewis OBE, P. Middleton, and B. Winkleman. 1980. The times atlas of the world: comprehensive edition. Time Books, London. 227 pp.

Foster, R.B., J.L. Carr, and A.B. Forsyth (eds.). The Tambopata-Candamo Reserved Zone of Southeastern Perú: a biological assessment. RAP Working Papers No. 6, Conservation International, Washington, D.C. 184pp.

Goulding, M. 1981. Man and fisheries on an Amazon frontier. Dr. W. Junk Publishers, Boston. 137 pp.

Hoorn, C. 1996. Miocene deposits in the Amazonian foreland basin. Science 273:122.

Killeen, T.J. 1998. Introduction. *In* Killeen, T.J., and T.S. Schulenberg (eds.). A biological assessment of Parque Nacional Noel Kempff Mercado, Bolivia. Pp 43-51. RAP Working Papers No. 10. Conservation International, Washington, D.C. 372 pp.

Marshall, L.G., and J.G. Lundberg. 1996. Miocene deposits in the Amazonian forelandbasin. Science 273: 123-124.

Räsänen, M.E., J.S. Salo, and R.J. Kalliola. 1987. Fluvial perturbance in the western Amazon basin: regulation by long-term sub-Andean tectonics. Science 238: 1398-1401.

Räsänen, M.E., A.M. Linna, J.C.R. Santos, and F.R. Negri. 1995. Late Miocene tidal deposits in the Amazonian foreland basin. Science 269: 386-390.

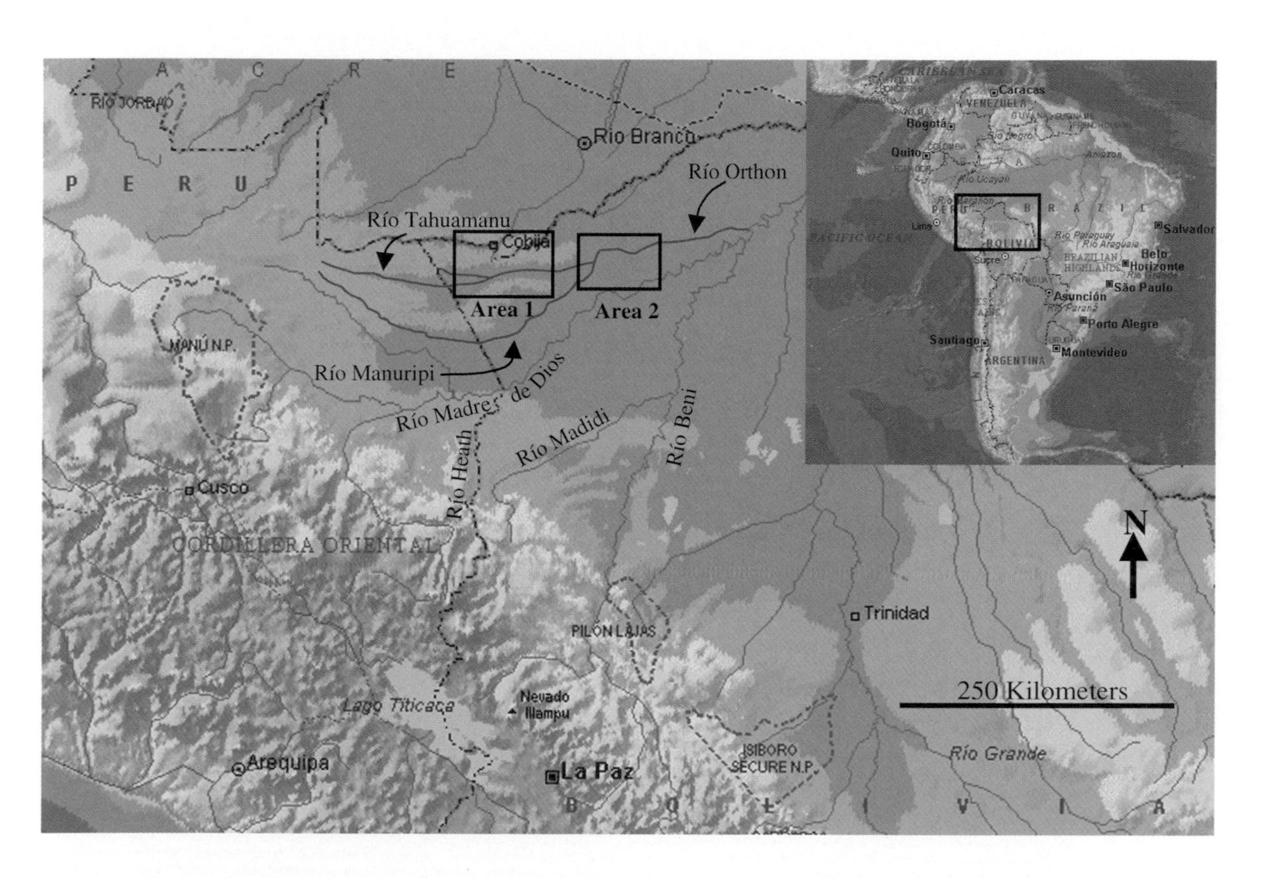

Map 1. Areas sampled during 1996 AquaRAP expedition to the Upper Río Orthon basin, Pando, Bolivia. Details of Area 1 are provided in Map 2, and details of Area 2 are given in Map 3.

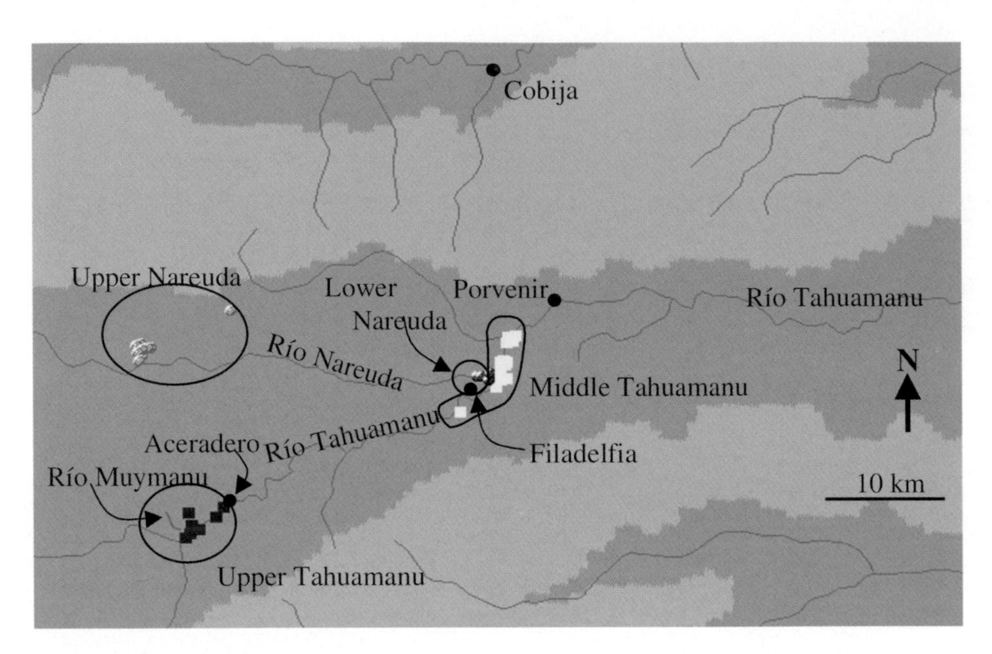

Map 2. Río Nareuda (upper = yellow circle; lower = gray circle) and Río Tahuamanu (upper = red squares; middle = yellow squares) localities sampled during the 1996 AquaRAP expedition to Bolivia.

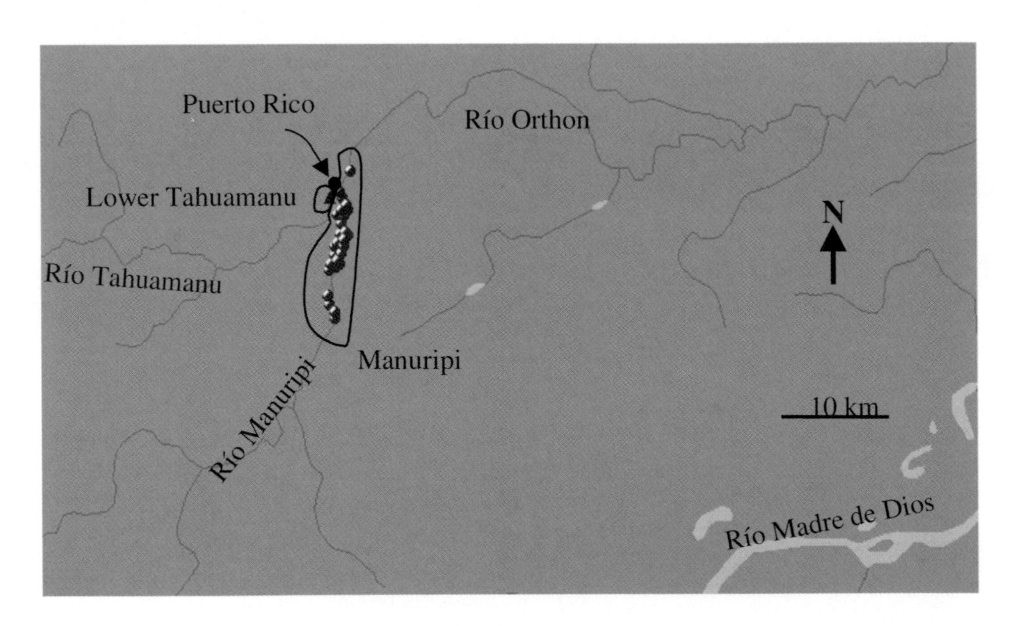

Map 3. Río Manuripi (red circle) and Lower Río Tahuamanu (purple triangle) localities sampled during the 1996 AquaRAP expedition to Bolivia.

T.Bert

Habitat conversion along the Upper Nareuda, Pando, Bolivia. Stage 1, Burning.

T.Bert

Habitat conversion along the Upper Nareuda, Pando, Bolivia. Stage 2, Cattle rearing.

T.Bert

Antonio Machado (front) and Barry Chernoff (rear) work the gill nets to collect fishes in the Río Manuripi, Pando, Bolivia.

T.Bert

Barry Chernoff and Naércio Menezes collect fish with a seine net.

K. Awbrey

AquaRAP Team members Barry Chernoff, Theresa Bert, Naércio Menezes, Hernán Ortega, and Roxana Coca, Pando, Bolivia.

T.Bert

A new species of pirahna, *Serraslmus* n. sp., from the Río Manuripi. Body length 35 cm.

K. Awbrey

AquaRAP team on a small oxbow lake, or "cocha," to the side of the Río Tahuamanu, Pando, Bolivia.

T.Bert

The armoured catfish, *Liposarcus disjunctivus*, from the Río Manuripi, Pando, Bolivia, commonly burrows into the sides of muddy riverbanks and eats insects, algae and detritus. Body length 42 cm.

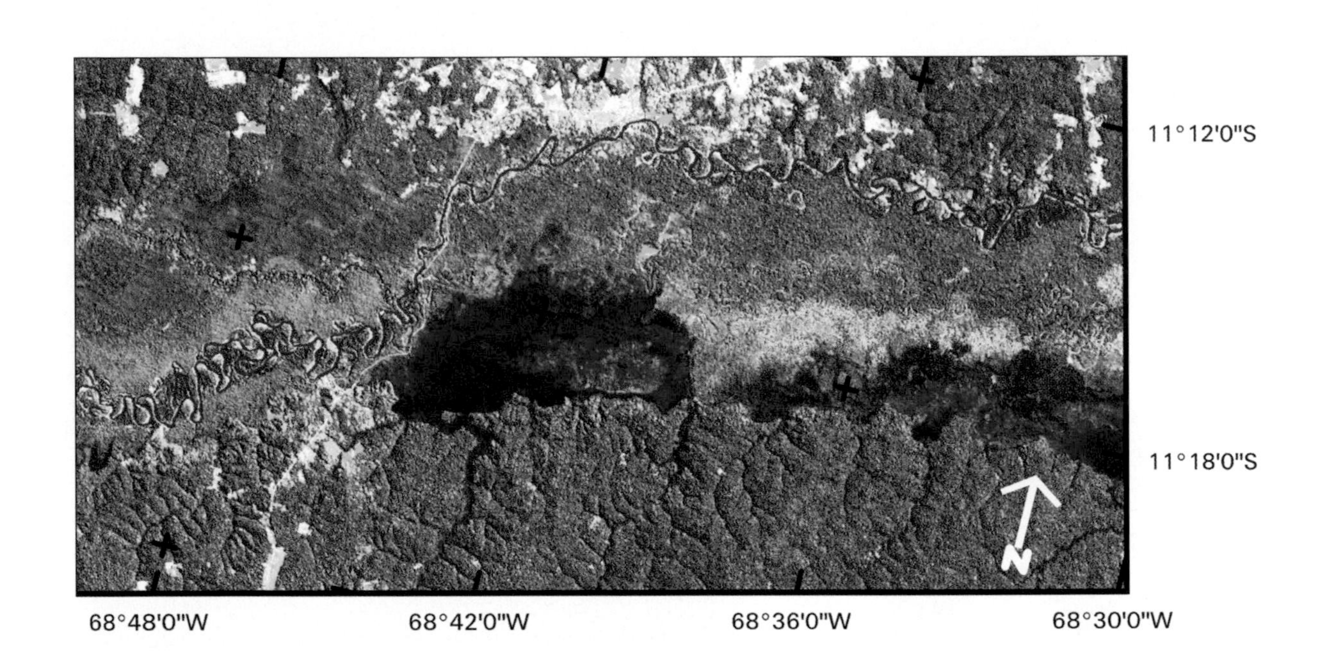

Satellite image of Middle Tahuamanu sub-basin within the Río Orthon basin, Pando, Bolivia. Image map prepared by the GIS lab at the Museo Noel Kempff Mercado using Lansat TM imagery provided by the NASA Lansat Pathfinder Project at the Department of Geography at the University of Maryland. Image number 002-068, June 1992.

CHAPTER 1

WATER QUALITY, ZOOPLANKTON AND MACROINVERTEBRATES OF THE RIO TAHUAMANU AND THE RIO NAREUDA

Francisco Antonio R. Barbosa, Fernando Villarte V., Juan Fernando Guerra Serrudo, Germana de Paula Castro Prates Renault, Paulina María Maia-Barbosa, Rosa María Menéndez, Marcos Callisto Faria Pereira, and Juliana de Abreu Vianna

ABSTRACT

In this report, we present the results of a rapid assessment of the water quality and the zooplanktonic and benthic macroinvertebrate communities at two regions in Pando, Bolivia during the period of September 4 to 20, 1996. Water quality was assessed at 22 sites. Waters in both regions were generally white, slightly acidic to neutral, and well oxygenated. The concentration of nutrients was medium, and temperatures ranged between 19 and 31° C. There were also some blackwaters as well as habitats where oxygen was depleted (e.g. shallow lakes). In general, water was of good quality and did not exhibit significant signs of contamination or eutrophication.

Zooplankton communities were assessed at 10 localities. A total of 120 taxa were identified. Rotifera had the highest proportion of species (44%), although Protozoa (40%) were also specious. Cladocera, Copepoda, Gastrotricha, Ostracoda, and Nematoda were also represented. Taxa could be classified as planktonic, semi-planktonic, or littoral/benthic. Most have cosmopolitan or tropical distributions.

Benthic macroinvertebrate communities were assessed at 8 localities. A total of 1,665 organisms were identified. Bivalvia were most numerous (27%), followed by Chironomidae, Heteroptera, and Oligochaeta. Gastropoda, Hirudinea, Decapoda, Arachnida, Odonata, Coleoptera, and Diptera were also represented. Bivalvia were most common in neutral to slightly alkaline waters. Among Chironomidae, 17 taxa were identified, and most of these are associated with the sediments. The benthic macroinvertebrate fauna showed a high density of predators, although herbivores were also collected. Based on Shannon diversity indices and work done in the Upper/Middle Río Paraguay, six sites were considered to have intermediate diversity and two sites had low diversity. It is hypothesized that no sites were categorized as high diversity because small taxa were not collected due to limitations in sampling methods.

INTRODUCTION

We present water quality analyses and the first assessments of zooplanktonic and benthic macroinvertebrate communities in two sub-basins, Upper Tahuamanu and Lower Nareuda/Middle Tahuamanu, in Pando, Bolivia during the period of September 4 to 20, 1996. The goal is to provide a basic physical and chemical characterization of the sampling sites and a preliminary view of the composition and distribution of the existing zooplanktonic and benthic biodiversity within these localities. An additional goal is to contribute to the improvement of the proposed AquaRAP protocols and their future utilization in other areas.

It must be emphasized that the presented results on zooplankton and benthic macroinvertebrates are restricted only to some of the sampling sites since some samples have been lost or were not satisfactorily preserved for microscope identification. Furthermore, the identification of the algal community is still being conducted, the results of which shall be published at a later date.

METHODS

At each site, the type of water, distance from the shore, depth, substrate type and vegetation were recorded. For the physico-chemical (water quality) characterization of the waters, the following variables were measured *in situ*: water temperature (°C), electrical conductivity (μS/cm), pH, and dissolved oxygen (mg/L and % saturation). The samples for the physico-chemical measurements were collected with a plastic jar lowered to 0.20 m depth.

Since we did not have a dredge, surber or corer to sample the benthic fauna, an insect net of 1 mm mesh size was used to sample 1 m^2 areas near the shore. For the plankton samplings at each site, 2 x 40 liters of sub-surface (approx. 0.20m depth) water were filtered through a 20 µm plankton net, fixed immediately after with 4% formaldehyde and examined under stereomicroscopy.

Map 2 shows a general view of the main drainages of the studied area, emphasizing the Río Tahuamanu. Localities and sites are described in detail in Tables 1.1 and 1.2, and in Appendix 1.

RESULTS AND DISCUSSION

Water Quality

In the Upper Tahuamanu region, the waters are generally those classified as whitewaters according to Sioli (1950). At this time, they showed temperatures ranging between 22.2°C and 31.2°C and pHs ranging between 5.46 and 7.90. The waters were well oxygenated (108.4-154.6% saturation), although low values (10-23% sat.) were also recorded, mainly in stagnated waters or small "curiches" along side the main rivers. Nutrient concentration was medium, as judged by the electrical conductivity values which ranged between 30.9 and 335.0 µS/cm.

The sites possessing riparian vegetation (e.g. site 1, Table 1.1, Appendix 1), rapids (e.g. sites 5 and 6), and some of the lakes with dense algae populations (sites 2, 4, and 7) are the ones showing well oxygenated waters despite the presence of high amounts of nutrients, as indicated by the electrical conductivity (e.g. site 5). On the other hand, the small, shallow and temporary lakes possessing high amounts of suspended fine particulate matter (sites 8, 9 and 10) show significant oxygen depletion, especially during the afternoon periods. This suggests the dominance of decomposition processes facilitated by high water temperatures. Considering the large amounts of plant material and the dark color of their waters, it is very likely that high amounts of humic substances are present, suggesting the low availability of inorganic nutrients. Table 1.1 summarizes the physico-chemical features for each station and their geographical positions in the Upper Tahuamanu sub-basin.

The Lower Nareuda/Middle Tahuamanu region is characterized mostly by whitewaters (Table 1.2), although black ones are present (e.g. site 11). Water temperature ranged between 19.7°C and 28.6°C, was slightly acidic to neutral (pHs between 5.71 and 7.62), and with oxygen levels normally above 80% saturation except those shallow temporary lakes where considerable oxygen depletion was recorded (30.7-39.5 % saturation). As in the previous sub-basin, the waters are of medium nutrient richness as judged by the electrical conductivity which ranged between 12.3 µS/cm (a small creek on the right margin of the Río Tahuamanu) and 159.2 µS/cm amidst central rapids in the same river. They are, in general, waters of good quality and did not exhibit any sign of significant contamination or eutrophication. Table 1.2 summarizes the physico-chemical features for each station and their geographical positions in the Lower Nareuda/Middle Tahuamanu sub-basin.

As in the Upper Tahuamanu, those areas possessing riparian vegetation and rapids (e.g. sites 16 and 21) are the ones showing high levels of oxygenation and are probably able to offer better conditions to maintain more diversified communities. In general, the quality of the waters in these two sub-basins are quite similar, although the number of sites with blackwaters is higher in the Lower Nareuda/Middle Tahuamanu sub-basin than in the Upper Tahuamanu.

Rapid Assessment of the Zooplankton Communities

The zooplankton records presented here were sampled from the following sites, covering lentic and lotic environments distributed along the Upper Tahuamanu and Lower Nareuda/Middle Tahuamanu sub-basins:

site 1: Río Tahuamanu (southeast and northwest margins)
site 2: Lake Aceradero
site 3: Río Muyamano
site 4: Marginal lake at Río Muyamano
site 5: Río Tahuamanu (before the mouth of Río Muyamano)
site 6: Mouth of Río Muyamano
site 7: Lake of Samaumas (limnetic and littoral zones)
site 8: Lake of Remanso
site 9: Lake Cañaveral
site 10: Curiche (northwest margin of Río Tahuamanu)

A total of 120 taxa were identified, among which 40% are Protozoa (tecamebas); 44% Rotifera; 7.5% Cladocera; 6% Copepoda; and 2.5% represented by Gastrotricha, Ostracoda, and Nematoda (Appendix 2).

Rotifera showed the highest species richness (53), mainly represented by the families Brachionidae (11 species), Lecanidae (11 species), and Trichocercidae (7 species). These were the major families recorded by Paggi (1990) in distinct environments within the floodplains of the middle Río Paraná and also by Brandorff et al. (1982) at Río Nhamundá in the Brazilian Amazon region. A predominance of Rotifera was also recorded by Hardy et al. (1984) within Lake Camaleão and by Sendacz (1991) in the floodplains of Río Acre.

Table 1.1. Physico-chemical characteristics of the waters from the Upper Tahuamanu sub-basin, Pando, Bolivia, September 1996.

Site Number	Date	Time	Location	Water temp.(°C)	pH	Conduct-Ivity (µS/cm)	Dis.Oxygen (mg/l)	Dis.Oxygen (% sat.)	Remarks
1	Sept, 4, 1996	9:00	11°26.699'S 69°0.509'W	28.4	7.43	270	8.85	114.3	Northwest margin; white water
1	Sept. 4, 1996	10:30	idem, idem	28.8	7.50	240	9.29	120.5	Southeast margin; riparian vegetation
2	Sept. 4, 1996	13:40	11°25.802'S 69°0.481'W	30.8	6.40	27.4	9.45	124.1	Lake
2	Sept. 4, 1996		11°25.872'S 69°0.189'W	29.1	6.46	30.9	10.0	126.7	Lake outlet
3	Sept. 5, 1996	9:50	11°27.225'S 69°1.810'W	27.8	6.85	78.9	9.47	124.2	Northwest margin
4	Sept. 5, 1996	10:25	11°27.197'S 69°1.834'W	28.7	6.72	122.8	8.47	111	Lake/river
5	Sept. 5, 1996	12:25	11°26.792'S 69°1.434'W	31.2	7.59	335	10.69	154.6	Mouth Río Tahuamanu; northwest margin
6	Sept. 5, 1996	13:30	11°26.839'S 69°1.468'W	28.3	6.97	83	10.41	134.6	Mouth Río Tahuamanu; southeast margin
7	Sept. 7, 1996	10:00	11°26.723'S 69°0.692'W	26.2	6.83	59.9	8.82	108.4	Lake at southeast margin Río Tahuamanu
8	Sept. 7, 1996	14:00	11°26.054'S 69°0.879'W	27.5	6.66	92.7	?	?	Lake at northwest margin Río Tahuamanu
9	Sept. 7, 1996	16:30	11°26.257'S 69°1.997'W	31.0	7.90	92.4	0.79	10.3	Lake at northwest margin Río Tahuamanu
10	Sept. 8, 1996	12:30	11°25.263'S 69°5.160'W	22.2	5.46	31	2.21	23.0	Curiche at northwest margin Río Tahuamanu

Table 1.2. Physico-chemical characteristics recorded at Lower Nareuda/Middle Tahuamanu sub-basin, Pando, Bolivia, in September 1996.

Site Number	Date	Time	Location	Water temp.(°C)	pH	Conduct-ivity (µS/cm)	Dis.Oxygen (mg/l)	Dis.Oxygen (% sat.)	Remarks
11	Sept. 10, 1996	11:00	11°18.336'S 68°45.861'W	22.8	6.05	40.2	3.48	39.5	Curiche at north margin Río Nareuda
12	Sept. 10, 1996	12:15	11°18.541'S 68°45.938'W	24.1	6.71	44.2	7.37	84.3	Río Nareuda, north margin
13	Sept. 11, 1996	10:00	11°18.525'S 68°45.965'W	19.7	6.00	40.6	3.08	31.6	Curiche at north margin Río Nareuda
14	Sept. 11, 1996	13:20	11°18.306'S 68°45.422'W	23.9	6.70	51.4	7.91	94.8	Río Nareuda, north margin
15	Sept. 12, 1996	11:00	11°20.5505'S 68°45.895'W	21.1	5.86	12.3	7.83	86.1	Arroyo north margin Río Tahuamanu
16	Sept. 12, 1996	13:30	11°20.567'S 68°46.061'W	26.7	7.46	175	9.31	116.6	Río Tahuamanu, north margin
17	Sept. 12, 1996	14:30	11°20.026'S 68°45.353'W	28.6	7.19	102.6	8.65	112	Lake at north margin Río Tahuamanu
18	Sept. 12, 1996	17:25	11°18.610'S 68°44.425'W	26.2	5.71	22.1	2.49	30.7	Lake at northwest margin Río Tahuamanu
19	Sept 13, 1996	10:20	11°16.273'S 68°44.338'W	22.5	6.20	17.5	8.05	87.0	Igarapé, southeast margin Río Tahuamanu
20	Sept 13, 1996	11:30	11°17.609'S 68°44.494'W	28.2	7.62	331	8.34	107.3	Lake at northwest margin Río Tahuamanu
21	Sept 13, 1996	12:30	11°18.273'S 68°44.482'W	27.4	7.38	159.2	7.51	88.0	Central rapids Río Tahuamanu
22	Sept 13, 1996	13:00	11°18.493'S 68°44.585'W	25.0	6.40	13.7	7.48	88.3	Arroyo, northwest margin Río Tahuamanu

The genus showing the highest number of taxa was *Lecane* (11 species), followed by *Trichocerca* (7) and *Brachionus* (5). According to Segers (1995), species of *Lecane* live mainly in littoral habitats, reaching highest diversity within tropical and subtropical waters. The genera showing highest occurrence among the sampled habitats were *Anuraeopsis, Trichocerca, Polyarthra, Brachionus, Keratella* and *Lecane*. Furthermore, the last three genera were also recorded as the most common by Vasquez et al. (1983) within two sections and a marginal lake of Río Orinoco, and according to Fernando (1980), *Keratella, Lecane* and *Brachionus* are typical components of the zooplankton fauna of tropical regions.

Among the recorded species of rotifers, 35 coincide with the ones listed by Koste et al. (1983, 1984) for Lake Camaleão (Brazilian Amazon), 31 coincide with the listed species recorded by Bonécker (1994) for the upper Río Paraná (Mato Grosso do Sul), 21 species recorded by Paggi (1990) for the floodplains of the middle Río Paraná, and 20 species recorded by Vasquez (1984) for Río Caroní (blackwaters) in Venezuela.

In general terms, the recorded rotifers show a cosmopolitan distribution. Examples are *Brachionus falcatus, Cephalodella gibba, Keratella cochlearis, Euchlanis dilatata, Lecane bulla, L. leontina, Plationus patulus,* and *Testudinella patina* among others. Some pantropical taxa were also recorded (e.g. *Anuraeopsis fissa, Brachionus caudatus* and *Keratella tropica)*, as well as some neotropical taxa, as represented by *Brachionus dolabratus* and *B. mirus*.

In relation to the diversity of habitats, species that inhabit the littoral/benthic zone were dominant, followed by the semi-planktonic species. The major planktonic species belong to the families Brachionidae and Synchaetidae. Among the Protozoa-Rhizopoda, 49 taxa were recorded. They were distributed among the families Difflugidae (19), Arcellidae (12), Centropyxidae (7), Eughyphidae (8), and Nebellidae (3). *Arcella* and *Difflugia* are the major genera with 12 and 14 species, respectively. These two genera were also recorded as dominant in the floodplains of the upper Río Paraná (Lansak Tôha, 1993).

The Cladocera were represented by 9 species distributed among 5 families, among which 3 are planktonic- Moinidae (1 species), Daphnidae (3 species), and 3 are typically from the littoral zone- Ilyocryptidae (1 species), Macrothricidae (1 species), and Chydoridae (3 species). Among these species, 5 were also reported by Green (1972) in lentic environments at the Río Suiá - Missú valley in the Pantanal wetlands. Furthermore, according to Sendacz (1991), *Ceriodaphnia cornuta* and *Moina minuta* are common species within the Brazilian Amazon region.

Copepoda were mainly represented by adults of Calanoida and Cyclopoida, and also nauplii and copepodits. A parasitic species (*Ergasilus* sp.) was also recorded in two of the lentic environments (Samaumas and Cañaveral).

Among the Cyclopoida, four species were identified (*Cryptocyclops brevifurca, Tropocyclops prasinus, Mesocyclops* sp. and *Microcyclops* sp.) and among the Calanoida, *Notodiaptomus* sp. Sendacz (1991). Also recorded was the occurrence of *Cryptocyclops brevifurca* in two lakes of the floodplain area of Río Acre. According to Dussart (1982), this species shows a wide distribution in South America, although it was reported by Sendacz (1991) as a rare species for the Amazon. Finally, a planktonic and cosmopolitan species of Cyclopoida, *Tropocyclops prasinus,* was recorded at site 1 (Río Tahuamanu). A subspecies (*T. prasinus meridionalis*) is restricted to South America according to Reid (1985).

Among the microcrustaceans (Copepoda, Cladocera, and Ostracoda), the highest richness was recorded at site 1 (Río Tahuamanu = 6 taxa), site 2 (Lake Aceradero = 10 taxa), and site 3 (Río Muyamano = 4 taxa). For total zooplankton, the highest richness was recorded at the right margin of the Río Tahuamanu (site 1 = 36 taxa), at Lake Aceradero (site 2 = 45 taxa), and at Lake Remanso (site 8 = 32 taxa). On the other hand, the poorest sites were Lake Cañaveral (site 9 = 11 taxa) and the curiche (site 10 = 12 taxa).

Except for sites 1 (northwest margin), 3 (Río Muyamano), 4 (marginal lake of the Río Muyamano), and 6 (mouth of Río Muyamano), rotifers were the predominant organisms, representing 40% (site 2) to 82% (site 9) of the total zooplankton community. Tecamoeba represented 0% (site 9) to 55% (site 6) of this community.

The most frequent species (present in at least 6 of the sampling sites) were *Polyarthra* sp., *Keratella tropica, Anuraeopsis fissa, Brachionus caudatus, Bdelloidea*, among the rotifers, juveniles of Copepoda, and *Arcella hemisphaerica* among the protozoa.

According to Jaccard index, there is a low similarity (27%) between the margins of the Río Tahuamanu (site 1), with only 14 taxa shared between northwest and southeast margins. Furthermore, a higher number of taxa was recorded at the right margin. This was probably due to the presence of a higher diversity of habitats and available food supply offered by the existing riparian vegetation at the southeast margin, as opposed to the presence of sand beaches at the northwest margin. In general terms, the studied sites showed the occurrence of planktonic, semi-planktonic, and littoral species of cosmopolitan and tropical distribution.

Rapid Assessment of the Benthic Macroinvertebrate Diversity

The sediment samples showed the presence of organic matter that can be divided into two distinct groups, as follows:

i) presence of plant debris (fragments of leaves, branches and seeds), clearly suggesting the contribution of allochthonous matter (sites 1, 2, 9, 13, and 20);
ii) dark organic sediment in advanced stages of decomposition, without the presence of stones or any larger particles (sites 8, 16, and 18).

A total of 1,665 benthic organisms were recorded from 8 sampling sites covering the two hydrographic basins (Appendices 3 and 4). The dominant groups were Bivalvia (27.3%), Chironomidae (23.7%), Heteroptera (18.1%), and Oligochaeta (10.3%). Lower abundances were also recorded for Gastropoda (*Biomphalaria* sp.), Hirudinea, Amphipoda (Gammaridae), Aracnida, Odonata, Coleoptera, and Diptera (Ceratopogonidae, Chaoboridae, Culicidae, Tabanidae, and Tipulidae). Among the Chironomidae larvae, 17 taxa were identified: *Ablabesmyia, Coelotanypus, Djalmabatista, Labrundinia, Asheum, Chironomus, Cladopelma, Fissimentum desiccatum, Goeldichironomus, Harnischia, Parachironomus, Polypedilum, Stenochironomus, Tribelos, Zavreliella,* and *Nimbocera paulensis* Trivinho-Strixino & Strixino (1991), along with some specimens of Tanytarsini. The majority are undescribed and endemic to South America (Fittkau & Reiss, 1973).

Most of the recorded Chironomidae taxa live within the sediments and water column in shallow environments (Nessimian, 1997), although some can reach the surface waters in search of prey and/or in order to escape unfavorable conditions such as low oxygen concentrations due to the high activity of microorganism decomposers of organic matter. The occurrence of larvae of *Fissimentum desiccatum*, (Cranston & Nolte 1996), a typical inhabitant of the potamal region but also occurring in areas with sandy and organic sediments with decomposing aquatic macrophytes, deserves special attention. According to Cranston & Nolte (1996), the larvae of *F. desiccatum* build galleries in the sediment and are able to resist dry periods, returning after rehydration.

In general, a higher abundance of Bivalvia was observed at sampling sites with waters showing neutral pH or slightly alkaline conditions, such as recorded for sites 1, 9, 16, and 20.

Among the recorded benthic macroinvertebrates, predators showed higher densities, particularly for Odonata, Heteroptera, Coleoptera, Ceratopogonidae, Chaoboridae, Tabanidae, and Tipulidae; along with some genera of Chironomidae (Tanypodinae), such as *Ablabesmyia, Coelotanypus, Djalmabatista,* and *Labrundinia.* This is a similar structure as the one recorded by Wong et al. (1998) in two Canadian lakes. Taking into consideration that carnivorous individuals would tend to be larger in size (such as Odonata nymphs), benthofagous fishes would be forced to feed upon smaller organisms which very often utilize feeding behaviors described as gathering-collectors, scrapers, or herbivores, such as exhibited by Ephemeroptera, Mollusca, and Chironomidae. This could be a reasonable explanation for the higher diversity of predators within the recorded communities when compared with other environments in the Amazon region (Fittkau, 1971; Reiss, 1977; Callisto & Esteves, 1996, 1998).

Besides predators, Chironomidae genera were also classified as follows: detritivores and/or miners (*Asheum, Cladopelma, Harnischia, Parachironomus, Stenochironomus, Tribelos* and *Polypedilum)*; collectors and shredders (*Chironomus* and *Polypedilum*); exclusively collectors (*Zavreliella*); collectors and filtrators (*Nimbocera paulensis* and Tanytarsini genera varia) (following Nessimian 1997 and Callisto & Esteves 1998).

Following the AquaRAP expedition to the upper/middle Río Paraguay (1997), Barbosa & Callisto (in press) proposed a tentative classification of the sampled sites by taking as a basis the values of Shannon diversity indices and the distinct types of habitats/microhabitats. In the present study, recorded diversity values could be classified as low to intermediate. The absence of sites with high diversity index could be a consequence of sampling constraints, namely the mesh size (1 mm). This would select for the larger taxa.

Using the same classification suggested for the Paraguayan waters, the present sites would then be categorized as follows:

a) Areas of intermediate diversity: $1.8 < H' < 3.0$, represented by sampling stations 2, 8, 13, 16, 18 and 20 which are lentic ecosystems with high autochthonous organic matter content in the sediment in distinct decomposition stages.

b) Areas of low diversity: with an $H' < 1.8$ and represented by sampling stations 1 and 9, located respectively in the Río Tahuamanu and Lake Cañaveral. Both areas have sediments rich in fine sand and clay and low content of allochthonous matter.

LITERATURE CITED

Barbosa, F.A.R., and M. Callisto. (*in press*). Rapid assessment of water quality and diversity of benthic macroinvertebrates in the upper and middle Paraguay River using the AquaRAP approach. Verh. Internat. Verein. Limnol.

Bonecker, C. C., F. A. Lansac-Tôha, and A. Staub. 1994. Qualitative study of rotifers in different environments of the high Parana River. Rev. Unimar, 16 (Supl. 3):1-16.

Brandorff, G-O, W. Koste, and N. N. Smirnov. 1982. The composition and structure of rotiferan and crustacean communities of the lower Río Nhamundá, Amazonas, Brasil. Studies on Neotropical Fauna and Environment 17: 69-121.

Callisto, M., and F.A. Esteves. 1996. Macroinvertebrados bentônicos em dois lagos amazônicos: lago Batata (um ecossistema impactado por rejeito de bauxita) e lago Mussurá (Brasil). Acta Limnologica Brasiliensia 8: 137-147.

Callisto, M., and F.A. Esteves. 1998. Categorização funcional dos macroinvertebrados bentônicos em quatro ecossistemas lóticos sob influência das atividades de uma mineração de bauxita na Amazônia Central (Brasil). Oecologia Brasiliensis 5: 223-234.

Cranston, P.S., and U. Nolte. 1996. *Fissimentum*, a new genus of drought-tolerant Chironomini (Diptera, Chironomidae) from the Americas and Australia. Ent. News 107(1): 1-15.

Dussart, B. H. 1982. Copépodes des Antilles françaises. Rev. Hydrobiol. Trop. 15(4): 313-324.

Fernando, C. H. 1980. The Freshwater Zooplankton of Sri Lanka, with a Discussion of Tropical Freswater Zooplankton Composition. Int. Revue ges. Hydrobiol. 65(1): 85-125.

Fittkau, E.J. 1971. Distribution and ecology of amazonian chironomids (Diptera). Can. Entomol. 103: 407-413.

Fittkau, E.J., and F. Reiss. 1973. Amazonische Tanytarsini (Chironomidae, Diptera) I. Die Ríopreto-Gruppe der Gattung Tanytarsus. Studies on the Neotropical Fauna 8: 1-16.

Green, J. 1972. Freshwater ecology in the Mato Grosso, Central Brazil. II. Associations of Cladocera in meander lakes of the Río Suiá Missú. J. Nat. Hist. 6: 215-227.

Hardy, E. R., B. Robertson, and W. Kost. 1984. About the relationship between the zooplankton and fluctuating water levels of Lago Camaleão, a Central Amazonian varzea lake. Amazoniana 9: 43-52.

Koste, W., and B. Robertson. 1983. Taxonomic studies of the Rotifera (Phylum Aschelminthes) from a Central Amazonian varzea lake, Lago Camaleão. Amazoniana 8: 225-254.

Koste, W., B. Robertson, and E. R. Hardy. 1984. Further taxonomical studies of the Rotifera from Lago Camaleão, a Central Amazonian varzea lake (Ilha de Marchantaria, Río Solimões, Amazonas, Brasil). Amazoniana 8: 555-576.

Lansac-Tôha, F. A., A. F. Lima, S. M. Thomaz, and M. C. Roberto. 1993. Zooplancton de uma planície de i nundação do Río Paraná. II. Variação sazonal e influência dos níveis fluviométricos sobre a comunidade. Acta Limnologica Brasiliensia 6: 42-55.

Nessimian, J.L. 1997. Categorização funcional de macroinvertebrados de um brejo de dunas no Estado do Río de Janeiro. Rev. Brasil. Biol. 57(1): 135-145.

Paggi, J. C., and S. José de Paggi. 1990. Zooplâncton de ambientes lóticos e lênticos do Río Paraná Médio. Acta Limnologica Brasiliensia 3: 685-719.

Reid, J. W. 1985. Chave de identificação e lista de referências bibliográficas para as espécies continentais sulamericanas de vida livre da ordem Cyclopoida (Crustacea - Copepoda). Bolm. Zool. Univ. S.P. 9: 17-143.

Reiss, F. 1977. Qualitative and quantitative investigations on the macrobenthic fauna of Central Amazon lakes. I. Lago Tupé, a black water lake on the lower Río Negro. Amazoniana 6(2): 203-235.

Segers, H. 1995. Guides to the identification of the microinvertebrates of the continental waters of the world. Rotifera. Volume 2: the Lecanidae (Monogononta). SPB Academic Publishing, The Hague, The Netherlands. 226 pp.

Sendacz, S., and S. de S. Melo Costa. 1991. Caracterização do zooplâncton do Río Acre e lagos Lua Nova, Novo Andirá e Amapá (Amazônia, Brasil). Rev. Brasil. Biol, 51(2): 463-470.

Trivinho-Strixino, S., and G. Strixino. 1991. Duas novas espécies de Nimbocera Reiss (Diptera, Chironomidae) do Estado de São Paulo, Brasil. Revta. Bras. Ent. 35(1): 173-178.

Vásquez, E. 1984. El zooplancton de la sección baja de un río de aguas negras (Río Caroni) y de un embalse hidroelectrico (Macagua I). Venezuela. Mem. Soc. Cienc. Nat. La Salle, 41: 109-130.

Vásquez, E., and L. Sánchez. 1983. Variación estacional del plancton en dos sectores del Río Orinoco. Mem. Soc. Cienc. Nat. La Salle 44: 11-34.

Wong, A.H.K., D.D. Williams, D.J. McQueen, E. Demers, and C.W. Ramcharan. 1998. Macroinvertebrate abundance in two lakes with contrasting fish communities. Arch. Hydrobiol. 141 (3): 283-302.

CHAPTER 2

DIVERSITY AND ABUNDANCE OF DECAPOD CRUSTACEANS IN THE RIO TAHUAMANU AND RIO MANURIPI BASINS

Célio Magalhães

ABSTRACT

The decapod crustacean fauna was assessed based on material collected by the two fish teams in 45 stations. Ten species of shrimps and crabs, representing three families and six genera, were found. Of these, six are new records for Bolivia. All the species are widely distributed throughout the Amazon basin or even the neotropics and are typically found in whitewater conditions. Several species already recorded from Bolivia were not found, and this may be due to inadequate sampling methods. There were only minor differences in decapod community structure among the sub-basins. There were specific microhabitat associations, with some species preferring aquatic vegetation (e.g. *Macrobrachium amazonicum*), others preferring submerged litter and dead trunks (e.g. *Macrobrachium brasiliense*), and others with no readily discernable preferences (e.g. *Macrobrachium jelskii*). Shrimps and crabs play many important roles in aquatic and terrestrial food chains. It is anticipated that they can be conserved as long as their habitats are not significantly modified.

INTRODUCTION

The limited knowledge about the composition and distribution of the freshwater shrimp and crab fauna from Bolivia comes mostly from collections undertaken in the Beni and Mamoré river basins. These rare and sporadic records are scattered throughout the literature (Holthuis, 1952, 1966; Bott, 1969; Rodríguez, 1992; and others). The AquaRAP expedition to the Upper Río Orthon basin would have been a good oportunity to improve this knowledge. Unfortunately, a team devoted specifically to this group of crustaceans could not join the expedition. Even so, the collections made by the ichthyology team yielded an appreciable amount of material that rendered valuable information on this group.

METHODS

The main habitats surveyed were river channel, small forest streams, and lakes. The samples were taken in sandy and muddy-sand beaches, floating and rooted vegetation, and areas with submerged dead leaves and trunks. The method of capture used was mainly 2 and 5 meter long seines and, eventually, trawl gear. All the samples were taken during the day. Three main areas were arbitrarily chosen, according to the origin of the samples: (1) "Upper Tahuamanu": samples from localities in the vicinity of the confluence of the Muyumanu and Tahuamanu rivers; (2) "Lower Nareuda/ Middle Tahuamanu": samples from small streams and localities in the vicinity of the confluence of the Nareuda and Tahuamanu rivers; (3) "Manuripi": samples from localities in the vicinity of the confluence of the Manuripi and Tahuamanu rivers. The specimens were identified using the descriptions in Holthuis (1952), Rodrígues (1992), Pereira (1993), and Magalhães and Türkay (1996 a,b), and comparisions with specimens from the collection of the Instituto Nacional de Pesquisas da Amazônia (INPA) , in Manaus, Brazil. The specimens are housed in the Museo Nacional de Historia Natural de Bolivia, La Paz, except for a few samples that were deposited in the Crustacean Collection of INPA, Manaus, Brazil.

RESULTS AND DISCUSSION

Ten species of shrimps and crabs were found, representing three families and six genera. A list of the species is presented in Table 2.1 and Appendix 5. The identity of two species remains doubtful. One, *Macrobrachium brasiliense*, has cryptic habits and is more active at night. Because of this, the collecting gear and the diurnal samples captured only juvenile and subadult specimens. For precise identification of this species, fully grown adult males should be examined. The other, *Valdivia* cf. *serrata*, is included in a genus showing a high inter- and intraspecific variability, and a safe identification of the specimens is dependent upon a careful revision of the group.

All these species are broadly distributed in the Amazon basin and have already been recorded from other localities in the Brazilian and Peruvian Amazon, or even in other basins like the Orinoco river basin. In the Amazon basin, the distribution of these species is usually related to the sediment-rich whitewater rivers, which form large areas that are periodically flooded (varzea). The occurrence of these species in the surveyed area would be expected as the Tahuamanu has some typical whitewater river features, such as high turbidity, pH near neutral, and high conductivity.

On the other hand, it is remarkable that some other species commonly found in such an environment were not captured. The presence of the shrimp *Euryrhynchus amazoniensis*, which occurs in the Central Amazon varzea and in the Madeira river basin (Tiefenbacher, 1978), would be expected, as well as *Dilocarcinus pagei*, a freshwater crab widely distributed in the floodplain areas of the Amazon and Paraguay/Paraná river basins (Rodríguez, 1992) and usually associated with the floating vegetation. Other species also widespread throughout the Amazon basin, or at least already recorded in the Bolivian territory of the Madeira river basin, were not collected as well. This is the case with *Dilocarcinus truncatus, Popianna argentiniana, Sylviocarcinus pictus, Valdivia camerani* and *Zilchipsis oronensis*, according to the records made by Magalhães and Türkay (1996b) and Rodríguez (1992). The absence of some, or even all, of these species among the material examined is most likely related to the fact that the appropriate methods for capture have not been used rather than that they do not occur in the surveyed area. Even so, some species were recorded for the first time in Bolivia, as it is the case with the shrimps *Macrobrachium depressimanum, M. brasiliense* and *Acetes paraguayensis*, and the crabs *Sylviocarcinus devillei, S. maldonadoensis* and *Valdivia* cf. *serrata*.

Table 2.1. List of the species of freshwater shrimps and crabs collected by the AquaRAP expedition to the Upper Río Orthon basin, Pando, Bolivia.

SPECIES	SAMPLING AREAS		
	Upper Río Tahuamanu	**Lower Río Nareuda/ Middle Río Tahuamanu**	**Río Manuripi**
Palaemonidae (Shrimps)			
Macrobrachium amazonicum	X		X
*Macrobrachium depressimanum**	X	X	X
Macrobrachium jelskii	X	X	X
*Macrabrachium brasiliense**		X	
Palaemonetes ivonicus	X	X	X
Sergestidae (Shrimps)			
*Acetes paraguayensis**		X	X
Trichodactylidae (Crabs)			
*Sylviocarcinus devillei**	X	X	
*Sylviocarcinus maldonadoensis**		X	
Valdivia cf. *serrata**	X	X	X
Zilchiopsis oronensis		X	

* First record for Bolivia

The difference in the composition of the species among the three main areas is hardly significant since most of the species were found in at least two or even all three of the areas. However, it is remarkable that in the Nareuda area, *Macrobrachium amazonicum* is absent and *M. brasiliense* is present. Even if the methodological problems mentioned above are taken into account, such a fact could be related to habitats. *Macrobrachium amazonicum* has always been collected in samples taken from floating aquatic vegetation, a habitat commonly found in and along the areas of the medium-to large-sized Muyumanu and Manuripi rivers. On the other hand, *M. brasiliense* was only found in environments with submerged litter and dead trunks, typical of small forest streams from the Nareuda area. Concerning the crabs, the presence or absence of the species in those areas could be mainly related to the use of non-appropriate methods for their capture.

Among the shrimps, there is a slight dominance of *Macrobrachium depressimanus*, both in terms of numbers of specimens and in terms of the number of samples in which it appears. This species was found both in rivers and in small forest streams, usually on sandy and muddy-sandy beaches or in places with submerged litter and dead trunks. Apparently, it is not associated with aquatic vegetation. It was not found in the few samples taken in the aquatic vegetation. In such an environment, the dominant species was *M. amazonicum*. Concerning *M. jelskii*, in spite of being less abundant than *M. depressimanum*, it seems to occur in the same habitats as the latter but was also collected in aquatic vegetation. Similar to the latter species, *Palaemonetes ivonicus* was found in the entire studied area, although its preferential habitat seems to be the medium - and large-sized rivers, usually found on beaches or floating vegetation. In the samples from the Nareuda area, it was collected in these habitats and not in the small forest streams, where submerged litter and dead trunks are more common. In regards to other species for which only a few specimens were collected, it is difficult to infer anything about their association with the habitats.

Human activities conferring some value to any of the shrimps and crabs species from the studied area could not be verified. However, ecologically this group is very important in the food chain of the aquatic environments. Shrimps and crabs play many roles, either as herbivores or as predators of various other invertebrate and vertebrate groups (Kensley and Walker, 1982; Walker, 1987, 1990; Schlüter and Salas, 1991), or even as prey for fishes (Goulding and Ferreira, 1984; Walker, 1987, 1990), birds (Beissinger et al., 1988), reptiles (Magnusson et al., 1984) and aquatic mammals (Benetton et al., 1988; Colares et al., 1996). If there is no commercial activity that could be of some threat to the natural populations of these species, they could be easily preserved as long as their habitats do not undergo any significant changes.

LITERATURE CITED

Beissinger, S.R., B.T. Thomas, and S.D. Strahl. 1988. Vocalizations, food habits, and nesting biology of the slender-billed kite with comparisons to the snail kite. Wilson Bulletin 100(4): 604-616.

Benetton, M.L.F.M., F.C.W. Rosas, and E.P. Colares. 1990. Aspectos do hábito alimentar da ariranha (*Pteronura brasiliensis*) na Amazônia brasileira. In: Programa y Resumenes, 4ª Reunión de Trabajo de Especialistas en Mamiferos Acuaticos de America del Sur, Valdivia, Chile, 12-15 Nov. 1990.

Bott, R. 1969. Die Süsswasserkrabben Süd-Amerikas undihre Stammesgeschichte. Eine Revision der Trichodactylidae und der Pseudothelphusidae östlich der Anden (Crustacea, Decapoda). Abhandlungen der Senckenbergischen Naturforschenden Gesellschaft 518: 1-94.

Colares, E.P., H.F. Waldemarin, and J.S. Said. 1996. Predação de crustáceos pela lontra (*Lutra longicaudis*) na região costeira do estado do Río Grande do Sul, Brasil. In: Resumos do III Encontro de Especialistas em Decapoda Brachyura, Río Grande, Brasil, 12-14 de dezembro de 1996. p. 22.

Goulding, M., and E. Ferreira. 1984. Shrimp-eating fishes and a case of prey switching in Amazon rivers. Revista Brasileira de Zoologia 2(3): 85-97.

Holthuis, L.B. 1952. A general revision of the Palaemonidae (Crustacea Decapoda Natantia) of the Americas. II. The subfamily Palaemonidae. Occasional Papers, Allan Hancock Foundation 12: 1-396.

Holthuis, L.B. 1966. A collection of freshwater prawns (Crustacea Decapoda, Palaemonidae) from Amazonia, Brazil, collected by Dr. G. Marlier. Bull. Inst. R. Sci. Nat. Belg. 42(10): 1-11.

Kensley, B., and I. Walker. 1982. Palaemonid shrimps from the Amazon basin, Brazil (Crustacea: Decapoda: Natantia). Smithsonian Contribution to Zoology 362: 1-28.

Magalhães, C., and M. Türkay. 1996a. Taxonomy of the Neotropical freshwater crab family Trichodactylidae. I. The generic system with description of some new genera (Crustacea: Decapoda: Brachyura). Senckenbergiana Biologica 75(1/2): 63-95.

Magalhães, C., and M. Türkay. 1996b. Taxonomy of the Neotropical freshwater crab family Trichodactylidae II. the genera *Forsteria, Melocarcinus, Sylviocarcinus,* and *Zilchiopsis* (Crustacea: Decapoda: Brachyura). Senckenbergiana Biologica 75(1/2): 97-130.

Magnusson, W.E., E.V. da Silva, and A.P. Lima. 1987. Diets of Amazonian crocodilians. Journal of Herpetology 21(2): 85-95.

Pereira, G. 1993. A description of a new species of *Macrobrachium* from Perú, and distributional records for *Macrobrachium brasiliense* (Heller) (Crustacea: Decapoda: Palaemonidae). Proceedings of the Biological Society of Washington 106(2): 339-345.

Rodríguez, G. 1992. The freshwater crabs of America. family Trichodactylidae and supplement to the family Pseudothelphusidae. Paris, Editions ORSTOM. 189 pp. (Collection Faune Tropicale, 31).

Schlüter, A., and A.W. Salas. 1991. Reproduction, tadpoles, and ecological aspects of three syntopic microhylid species from Perú (Amplibia: Microhylidae). Stuttgarter Beitr. Naturk., Ser. A. nº 458: 1-17.

Tiefenbacher, L. 1978. Zur Systematik und Verbreitung der Euryrhynchinae (Decapoda, Natantia, Palaemonidae). Crustaceana 35(2): 178-189.

Walker, I. 1987. The biology of streams as part of Amazonian forest ecology. Experientia 43: 279-287.

Walker, I. 1990. Ecologia e biologia dos igapós e igarapés. Ciência Hoje 11(64): 45-53.

CHAPTER 3

FISHES OF THE RIOS TAHUAMANU, MANURIPI AND NAREUDA, DEPTO. PANDO, BOLIVIA: DIVERSITY, DISTRIBUTION, CRITICAL HABITATS AND ECONOMIC VALUE

Barry Chernoff, Philip W. Willink, Jaime Sarmiento, Soraya Barrera, Antonio Machado-Allison, Naércio Menezes, and Hernán Ortega

ABSTRACT

Two teams of ichthyologists surveyed the freshwater fishes of the Río Tahuamanu, Río Manuripi and Río Nareuda, Pando, Bolivia at 85 collecting stations for 17 days. They captured 313 species of which 87 are new records for Bolivia. These records bring the total fish fauna of Bolivia to 641 species and for the Bolivian Amazon to 501 species. This small region in northeastern Bolivia contains 63% and 49% of all the species known to inhabit the Bolivian Amazon and Bolivia, respectively. This region is potentially a hotspot of fish biodiversity. The habitats on which these fishes depend is discussed, from which we recommend the preservation of key flooded areas, main river channels, and small streams or tributaries passing through terra firme habitats. A large percentage of the fauna have high economic value as food or ornamental fishes. Stocks are in need of immediate evaluation, and the exportation of food fishes needs to be regulated as soon as possible.

INTRODUCTION

The forested lowlands of most amazonian tributaries remain poorly explored. Lists of freshwater fishes do not exist for more than one or two of the major arms of the Amazon, though Welcomme (1990) lists 389 species for the Río Madeira. Goulding (1981) produced a remarkable work on the fishes and fisheries of the Río Madeira, but provided no list of species. In Bolivia there have been attempts to bring together information for amazonian streams (e.g., Lauzanne et al., 1991). More recently, a well documented fauna from the Río Gaupore/Itenez was provided by Sarmiento (1998). Lauzanne et al. (1991) recorded 389 species of freshwater fishes from the Bolivian Amazon to which Sarmiento (1998) added an additional 21 species, bringing the total to 410 species.

This work reports upon the ichthyological results of the first AquaRAP survey carried out in a relatively small area of the upper Río Orthon. Two teams of ichthyologists surveyed aquatic habitats in the Ríos Tahuamanu and Manuripi and their tributaries over 17 days in 1996. Amazingly, 313 species of fishes were recorded. This value includes a spectacular number of new records for Bolivia and several species new to science, including a remarkable new piranha in the genus *Serrasalmus*.

The purpose of this report is to: (i) make known the general ichthyological results, including a breakdown by sub-basin; (ii) compare the diversity with those of other regions and discuss any distributional aspects of the data; (iii) discuss the economic and conservation importance of the results; (iv) discuss specific threats to the biodiversity; and (v) discuss recommendations for conservation and future research. In a companion paper (Chapter 5) , specific analyses of distributions within the study region and among macrohabitats and water type will be presented along with specific management recommendations.

METHODS

Two fish teams, each containing three to four members, made collections from 4 - 21 September 1996. The two teams usually worked apart and were camped in different sections of the Upper Río Tahuamanu for the first five days. Eighty-five collection stations were made, each receiving a unique, sequential field number (Appendix 8). The field stations were enumerated separately for each group, identified as P1 and P2. At each station, longitude and latitudes were obtained from hand-held GPS units that had been calibrated and checked at the airstrip in Cobija, Pando, Bolivia for which we have accurate coordinates. Obtaining geographic coordinates in the field was not always possible because of forest cover and position of satellites.

At each field station a number of ecological variables describing the habitat were recorded. These included descriptions of the shore, substrate, type of habitat (e.g., river, lake, flooded area, etc.) as well as the water type. The classification of water type (black, white, turbid) was checked with the results obtained by the limnology group.

Fishes were collected using a variety of nets and netting techniques. Each group was equipped with seines (5m x 2m x 1.25 cm, 5m x 2m x 0.63 cm, 1.3m x 0.7m x 0.37 cm), dip nets and experimental gill nets (40m x 2m, monofilament, with five 8m panels, mesh size from 1.25 cm to 6.25 cm). Team 2 pulled an otter trawl (mouth 3m wide) with two 15 kg doors where the depth of the water was >2m over sandy or muddy stretches in a manner modified from that described by López-Rojas et al. (1984). Additionally, one of the river pilots threw a 2m cast net in some deeper lakes or cochas.

Fishes were preserved in buffered 10% formalin solution. All specimens captured at the same place and time were maintained separately from all other collected specimens. Larger specimens were tagged individually using fine wire and punched cardboard tags and either placed in large liquidpacks containing formalin with other specimens or were skeletonized, soaked in 40% isopropanol and dried. All material was wrapped and shipped to Chicago for sorting, identification and enumeration in the Division of Fishes, Department of Zoology, Field Museum (FMNH). Fifty percent of the specimens are housed in the Museum of Natural History, La Paz, Bolivia; the remaining specimens were shared among the participating institutions: FMNH; Museu Zoologia do Universidade do São Paulo, Brazil; Museo de Historia Natural San Marcos, Perú, and Museo Biologica de la Universidad Central de Venezuela.

The identifications were made in a careful but relatively rapid fashion. General works such as Eigenmann and Myers (1929) or Gery (1977) were used, but preference was always given to systematic revisions (e.g., Vari, 1992; Mago-Leccia, 1994) and recent species descriptions (Stewart, 1985) if available. In many cases specimens were compared to types or historic material referenced in the literature and housed at FMNH. However, identification to the level of species or even genus was not always possible. To do so would represent a less than scholarly approach to the taxonomy. Instead we rely upon morphospecies – the number of distinguishable entities present in our samples. This bears the assumption that such discernable entities or morphospecies are putative taxonomic entities (i.e., species). We were careful to check for sexual and ontogenetic differences. All specimens were examined critically and identified to their lowest taxonomic level (Appendix 6).

Another issue inheres in the appropriate selection of taxa across lists in order to judge new geographic records (Appendix 7). We chose a conservative approach and did not include all of the taxa that we had collected. We eliminated from comparison those taxa whose identifications were ambiguous or unknown in our list AND in published lists. So for example, if *Hemigrammus* sp. or *Hemigrammus* sp. 1 occurred in both lists, it was not counted as a similarity or a difference because there is no way to ascertain that the taxonomic designations represent the same biological entity. However, *Gephyrocharax* sp. occurs in our list but only *G. chapare* is reported by either Lauzanne et al. (1991) or Sarmiento (1998). In this case we count the *Gephyrocharax* sp. as a new record because we compared our material to *G. chapare,* and it is different. The possible error in this latter case is identical to the possible errors in a list containing misidentified species bearing specific epithets or not.

Collecting Stations

Eighty-five collections were taken in the Tahuamanu and Manuripi river basins from the border with Peru downstream to Puerto Rico (Maps 1 - 3). The entire region was divided into the following six subregions: Upper, Middle and Lower Tahuamanu; Upper and Lower Nareuda; and the Manuripi. The Río Nareuda is a major tributary of the Río Tahuamanu.

The gear effort used to sample the 85 field stations were as follows: seine –73; trawl – 6; gill net – 5; cast net –2. This adds up to 86 because one station, P2-04, included both gill net and seine collections. The gill nets were usually set for several days. Because no striking differences were noted for day and night samples within the gill nets, they were recorded as single stations. Due to low water conditions during the time of the field sampling, trawling was difficult because motors had only limited use in the Upper Tahuamanu and could not be used in the Río Nareuda except for in the vicinity of its confluence with the Río Tahuamanu.

During the first portion of the expedition, Team 1 worked in the Upper Tahuamanu and also in the lower end of the Río Muymanu, whereas Team 2 collected in the Upper Nareuda as well as in a number of small streams (garapes) that drained independently into the Tahuamanu. The Upper Tahuamanu and the Upper Nareuda systems as well as their tributaries were surrounded by terra firme.

For the middle portion of the expedition, Team 1 focused upon the Lower Nareuda down to its mouth in the Río Tahuamanu. Team 2 collected upstream and downstream from this confluence to just below the village of Filadelfia; this region is referred to as Middle Tahuamanu. Conditions in the Middle Tahuamanu were such that trawling was accomplished successfully.

For the last period both teams camped on the Río Manuripi upstream from Puerto Rico. Though both groups worked independently, they covered the same territory in the Río Manuripi as well as in the Lower Tahuamanu, just above its mouth in the Río Manuripi. River conditions permitted trawling.

RESULTS AND DISCUSSION

Diversity and Distribution: General

The biodiversity of fishes was unexpectedly spectacular. A total of 313 species were captured and identified. The fishes (Appendix 6) included members of all trophic or activity groups, ornamentals (e.g., *Abramites hypselonotus*), food fishes (e.g., *Pseudoplatystoma fasciatum*), as well as miniatures (*Scoloplax* cf. *dicra*, <20 mm SL) and large fishes (e.g., *Prochilodus* cf. *nigricans, Doras* cf. *carinatus*, >200 mm SL).

Together Lauzanne et al. (1991) and Sarmiento (1998) record 410 species from all rivers within the Bolivian Amazon. The 313 species of fishes, therefore, represents a fauna greater than 76% of the number of species previously reported from all other Amazonian tributaries within Bolivia (Appendix 7). The number of fishes discovered in a relatively small section of the Tahuamanu and Manuripi rivers is more than three times that reported for the entire Beni-Madre de Dios basin (n=101, Lauzanne et al., 1991), more than 1.3 times that reported for the Río Guapore/ Itenez (n=246, Sarmiento, 1998) and almost equal to (96%) that found over the entire Río Mamore basin (n=327, Lauzanne et al., 1991). Furthermore, Santos et al. (1984) reported 300 species from the lower Río Tocantins of Brazil; Stewart et al. (1987) reported 473 species in the Napo River Basin of Ecuador; and Goulding et al. (1988) reported 450+ species from the Río Negro basin of Brazil. In each of these latter cases, the areas surveyed exceed vastly that of the Tahuamanu-Manuripi region sampled in the rapid assessment.

It is important to consider that the number of species reported for both the Guapore/Itenez and Mamore drainages included headwater habitats that usually contain fishes with restricted distributions. Even in the Upper Tahuamanu and Upper Nareuda, there were few habitats, perhaps the two garapes, that could correspond to headwater-like conditions.

Because of the paucity of knowledge concerning the distributions of the freshwater fishes of South America, it is difficult to state with any certainty the degree of endemism represented in AquaRAP samples from the upper Río Orthon. However, we have apparently uncovered not only a region with high biodiversity, but a region with an exceptional number of new records for Bolivia (Appendix 7). Using the conservative approach discussed above, we document 87 species not previously recorded from Bolivia. Of the 87 species, 45 are new to science (e.g., the new characid, *Chrysobrycon* sp. 1) or species with some questions associated with their exact name (e.g. *Anchoviella* cf. *carrikeri*). The newly documented fauna increases the total number of species inhabiting the Bolivian Amazon by about 21% to 501 and increases the total for all of Bolivia (using the previous total in Sarmiento, 1998) 16% to 641 species.

The significance of the Tahuamanu and Manuripi rivers for the Bolivian ichthyofauna is now clear. This limited region contains 63% of all fish species known from the Bolivian Amazon and 49% of all fishes known from Bolivia. Furthermore, the 87 species found in the Tahuamanu and Manuripi rivers that are not yet known elsewhere in Bolivia represent within-country "endemism" values of 18% and 14% relative to the ichthyofaunas for the Bolivian Amazon and Bolivia, respectively.

Based upon the relatively sparse information published for the ichthyofauna of Bolivia, of regions with more than 50 species, only Lake Titicaca has a higher within-country "endemism" percentage. No other region within Bolivia is currently known to contain as high a percent of the amazonian or total country fauna. The impressive values reported for the Río Guapore/Itenez basin within the Parque Nacional Noel Kempff reported upon by Sarmiento (1998) must now be adjusted downwards due to the new species totals. The adjusted values are well below those of the Tahuamanu and Manuripi rivers.

Potentially, the impressive nature of the Tahuamanu-Manuripi ichthyofauna could be tempered if, in fact, it was representative of a more widespread fauna within Bolivia. The most obvious possibility is that we used trawls to sample the bottom communities for the first time within Bolivian freshwaters. If the 87 novel records were largely due to trawl samples, then the lists would not really be comparable. That is, we might expect to find similar bottom communities in other regions; the uniqueness of the Tahuamanu-Manuripi region would diminish even though its overall diversity would remain exceptional. However, this was not observed. The trawls captured 53 species (17% of the total), but only 15 species were captured in trawls exclusively. Of these, 10 were new records for Bolivia. Thus, 77 of the 87 species newly reported for Bolivia were captured by traditional means used commonly in ichthyological sampling.

We have no doubt that more careful sampling as exemplified in the Guapore/Itenez by Sarmiento (1998) and in the Tahuamanu-Manuripi by the AquaRAP team will lead to increasing the number of taxa found within Bolivia and will increase knowledge of the distribution patterns of the fishes. Nonetheless, it seems unlikely that the distinctive character of the Tahuamanu-Manuripi fauna will lose its uniqueness.

In fact with continued sampling, the ichthyofauna of the Tahuamanu-Manuripi region will continue to rise and the number of species new to Bolivia should also increase. We base this estimate upon the species accumulation curves (Fig. 3.1). After 15 days of sampling, the rate of accumulation of species new to the expedition had not diminished; the graph displays no asymptote (Fig. 3.1). During the last six days, even with both groups working the same region of the Manuripi, we increased the known fauna by 63 species. Species were being added at an average rate of 10.5 species per day.

The total capture and accumulation rates were remarkably similar for each of the two collecting teams (Fig. 3.1). Each group captured more than 200 species. By the end of the sampling period, the species collected by the groups differed as much as 50%, which maintained the steep slope of the total accumulation curve. Analyses of these data argue strongly that continued collecting will increase the size of the fauna known from the Tahuamanu-Manuripi region. Because the species that represent new records for Bolivia comprise more than 29% of all species captured, we expect that additional collecting will continue to uncover new records for the country as well. These data exemplify that the Upper Orthon basin is extremely diverse and may prove more diverse than even the Río Mamore basin, which has received vastly more collecting effort.

The biogeographic relationships of the Tahuamanu-Manuripi region is difficult to ascertain, again because of how little we know in general. There does seem to be a mix of taxa representing three distinct distributional elements. The first comprises widespread amazonian lowland species from the north and the east of the Madeira basin: e.g., *Serrasalmus rhombeus*, *Schizodon fasciatum*, *Anodus elongatus*, *Microschemobrycon geisleri*, *Cetopsorhamdia fantasia*, *Pimelodus* cf. *altipinnis*, *Tatia altae*, *Electrophorus electricus*, and *Rhabdolichops caviceps*. The second incorporates species found in black waters of the Guapore/Itenez system that derive from the Brazilian Shield: e.g., *Hypopygus lepturus*, *Carnegiella strigata*, *Aphyocharax alburnus*, *Hemigrammus* cf. *unilineatus*, *Pyrrhulina australe*, *Nannostomus trifasciatus*, *Potamorhina latior*, *Tatia aulopygia*, *Corydoras hastatus*, *Cichla monoculus*, and *Aequidens* cf. *tetramerus*. The third describes those species in the small garapes that are typical for headwater habitats: e.g., *Chrysobrycon* spp., *Brachychalcinus copei*, *Bryconamericus* cf. *caucanus*, *Creagrutus* sp., *Cyphocharax spiluropsis*, *Piabucus melanostomus*, *Tyttocharax tambopatensis*, *Corydoras trilineatus*, *Imparfinis stictonotus*, and *Otocinclus mariae*.

The discussions above show the nature of the uniqueness of the ichthyofauna of the Tahuamanu-Manuripi river basins. A relatively small region contains 63% of all the freshwater species known from the Bolivian Amazon. The fauna comprises 87 species that have never been recorded from Bolivia – a feature which does not seem to be an artifact of sampling or collecting methods. The region contains an assemblage of fishes that may uniquely combine widespread amazonian species, with blackwater Brazilian shield species in addition to headwater species. Because we never reached the asymptote of the species accumulation curve, the number of species represented in this region is predictably larger than we can document at present. Furthermore, because the number of new records comprises 29% of the fauna of this region, it is reasonable to expect new records as well.

All of this leads to the conclusion that within Bolivia, the upper Río Orthon basin must be considered as a potential hot spot for the biodiversity of freshwater fishes. Conservation efforts are needed to preserve the unique character of this fish fauna. At a secondary level, more field studies are needed to finish documenting the ichthyofauna of this amazing region.

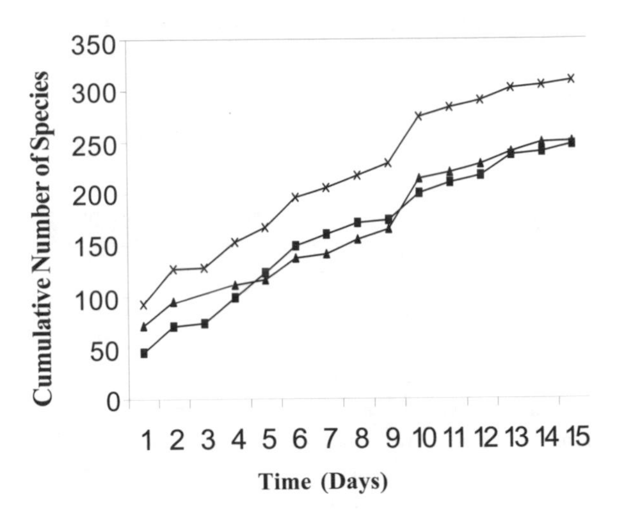

Figure 3.1. Species accumulation curves for fishes collected in the Ríos Tahuamanu, Nareuda, and Manuripi, Pando, Bolivia, 4 - 21 September, 1996. Symbols: Group 1 (squares), Group 2 (triangles), and combined (X).

Economic Importance

Many of the fishes found in the Upper Río Orthon basin are valuable as commercial resources for food and for the ornamental fish industry. We present a short discussion in order to stimulate immediate research into the potential of these resources to provide an economic alternative to other activities that damage the environmental character of the region.

During the AquaRAP we employed local fisherman who were subsistence fisherman throughout the year and who fished commercially during the seasonal migrations. We also witnessed truck-side fish sales in Cobija and in Filadelfia and saw the following species for sale: *Pygocentrus nattereri*, *Pseudoplatystoma fasciatum*, *Prochilodus nigricans*, *Curimata* spp., *Hydrolycus pectoralis*, *Plagioscion squamosissimus*, *Mylossoma duriventris*, and *Myleus* sp. We were told by the fishermen that the fishes of highest commercial value were the surubí, *Pseudoplatystoma*, and the serrasalmines, including both the piranhas and the pacus. In our samples, we caught many other species that are either consumed for subsistence or caught for sales according to our guides. That list included the following: *Anodus elongatus, Cichla* cf. *monoculus, Cochliodon cochliodon, Crenicichla* spp., *Duopalatinus* sp., *Hemisorubim platyrhynchos, Hoplias malabaricus, Megalechis thoracatus, Hypostomus* spp., *Leiarius marmoratus, Leporinus* spp., *Myleus* sp., *Pimelodus* spp., *Pristobrycon* sp., *Rhamdia* sp., *Schizodon fasciatum, Sorubim lima,* and *Triportheus angulatus.* In that list the "spp." refers to a number of species within the genus that are eaten. Interestingly, the following armored catfishes are consumed in the Río Madeira basin of Brazil (Goulding, 1981) but were rejected as a source of food by our fishermen: *Doras* cf. *carinatus, Pseudodoras niger*, and *Liposarcus disjunctivus.*

Our discussions with local fisherman including others in Puerto Rico indicated that the commercial food fisheries are burgeoning in the Tahuamanu and Manuripi rivers. It was not possible to obtain from our fishermen or from the truck-side sellers the annual totals for weight or value of the catch. However, the notion of increasing annual landings would not seem to make sense relative to the local populations that we encountered and interviewed for two reasons: i) the population of Pando, while increasing, is doing so slowly; and ii) the Bolivian residents have largely settled the region from cattle rearing areas or from La Paz and do not have strong traditions of fish consumption. Most of their demand is for premium species, tiger catfishes and pacus, species that are delicate in flavor. Apparently, much of the catch is exported across the border to Brazil, and this exportation appears to be unregulated. Sarmiento (1998) also noted a similar pattern of unregulated exportation of several tons of fish per month with a slightly increasing demand from populations living along the Río Itenez. The mere fact that catches are not being monitored is reason for concern.

Peres and Terbourgh (1995) , Goulding (1980, 1981) and Goulding et al. (1988) document not only the importance of rivers in structuring human settlements throughout Amazonia but also the increasing dependence of humans on aquatic resources for sustenance. At this time we cannot document the size of species-specific harvests that are sustainable for the future. There are no accurate data on the nature of fish migration into the Tahuamanu-Manuripi region. Many of the commercially important species such as the curimatids and prochilodontids move out of tributaries and forest habitats into the main rivers to spawn at the beginning of the flooding cycle (Goulding, 1981). Many of the larger catfishes migrate upstream to spawn and apparently ascend the cataracts in the upper Madeira into Bolivian waters (Goulding, 1981). The only relevant data that we collected was that the abundance of the most favored species was relatively low (Machado-Allison et al., 1999). Even though it was the dry season and at relatively low water, there were still many suitable habitats for these species. The gill net samples did not yield the quantity of individuals for the commercially important species that would indicate bountiful populations, even though our fishermen guided us to their favorite areas (with monetary exchange for the catch). In our experience from similar habitats in other regions of the Amazon and Orinoco river basins where fish populations are apparently healthy, the commercially important species are reasonably well represented in gill net samples even in the dry seasons. We only suggest caution. And we recommend that the stocks be surveyed immediately both within the Upper Río Orthon as well as downstream toward the Río Madeira. Statements such as that by Walters et al. (1982) advocating across - the - board increase in fisheries are premature in advance of the data on native stocks.

Economic potential also exists for the establishment of a harvest-based ornamental fishery. The extreme number of species, the number of cochas and inundated flooded habitats makes the Tahuamanu-Manuripi regions especially attractive. These habitats are easily collected while serving as natural critical rearing habitats for ornamental fishes. The ornamental fishes ranged from the common (e.g., *Moenkhausia sanctaefilomenae, Hemigrammus ocellifer*) to ornamentals that are more highly prized, including the following: *Astronotus crassipinnis, Apistogramma* spp., *Aequidens* spp., *Satanoperca* cf. *acuticeps, Eigenmannia* spp., *Apteronotus albifrons, Gymnotus carapo, Hypopygus lepturus, Heptapterus longior, Imparfinis stictonotus, Microglanis* sp., *Brachyrhamdia marthae, Cheirocerus eques, Scoloplax* cf. *dicra, Peckoltia arenaria, Parotocinclus* sp., *Otocinclus mariae, Hypostomus* spp., *Ancistrus* spp., *Agamyxis pectinifrons, Acanthodoras cataphractus, Brochis splendens, Corydoras* spp., *Bunocephalus* spp., *Dysichthys* spp., *Nannostomus trifasciatus, Pyrrhulina* spp., *Carnegiella* spp., *Tyttocharax* spp., *Poptella compressa, Phenacogaster* spp., *Prionobrama filigera, Metynnis luna, Paragoniates alburnus, Iguanodectes spilurus, Hemigrammus* spp., *Hyphessobrycon* spp. *Chrysobrycon* spp., *Aphyocharax* spp., *Leporinus* spp., *Cynolebias* spp., and *Pterolebias* spp.

The most important areas within the region surveyed for the ornamental fishes are the Upper Río Nareuda and the Río Manuripi (Appendix 6). In some cases, trapped interior flooded forest lakes (e.g., Appendix 8: B96-P2-42, B96-P2-44) can be harvested entirely because these ephemeral environments either dry completely or become anoxic. However, in the more permanent habitats the reproductive and population biologies of the ornamental fishes must be studied in order to support a sustainable enterprise.

Critical Habitats

We identified a number of critical habitats that are required for continued survival of freshwater fishes and maintenance of the spectacular biodiversity. These are the same habitats that support the growth and reproduction of economically valuable species.

The habitats are described in detail in the companion paper (Machado Allison et al., 1999). However, they fall into three main classes: i) flooded areas; ii) small tributaries; and iii) main channels. The flooded areas comprise the most critical and highly endangered areas, including the varzea (flooded forest), cochas, swamps, forest lakes, etc. These areas provide nursery grounds for perhaps 66% of the species that we captured (Goulding, 1980, 1981; Lowe-McConnell, 1987). Furthermore, many species, including the pacus, feed on the fruits and nuts, including the Brazil nuts, dropped by the plants into the water (see Goulding, 1980, 1981). Goulding (1981) characterizes the Brazilian portion of the Río Madeira as having a relatively narrow flood plain and margin. This is certainly true for the Ríos Tahuamanu, Muymanu, Nareuda and Manuripi in the areas that we surveyed. Given such a narrow flood plain, there is little buffer between logging and ranching activities and these critical flood zones. Extended development from Puerto Rico, Filadelfia and Aserradero threaten this flood plain zone.

In the Upper Tahuamanu and Upper Nareuda there were many smaller streams (garapes) and tributaries with both black and white water conditions. These smaller habitats are highly threatened by deforestation and cattle ranching. A number of the garapes crossing the main road and on cattle ranches are completely denuded of riparian vegetation. Furthermore, one team walked through dense forests into a number of forest streams and neither captured nor saw any fishes, crustaceans or aquatic insects in crystal clear water with sand bottoms. These streams were downstream from a ranch that was recently cleared and burned. We could only speculate that ashes, which are toxic to aquatic organisms (N. Menezes, pers. comm.), or pesticides poisoned these streams.

These upper areas are highly diverse. We captured 168 species in the Upper Tahuamanu and the Upper Nareuda, slightly more than half of all the species we discovered (Appendix 6). Seventy-one species were found to inhabit both the Upper Tahuamanu and Upper Nareuda, but 30 species were found only in these upper regions. These upper regions contain habitats that are similar to headwater areas with a unique and diverse fauna. These are among the most threatened habitats due to logging, deforestation and ranching.

The main river channels are not habitats that are usually focused upon. However, as pointed out above, the principal channels also contain a diverse fauna with a number of unique or rare elements (e.g., *Cetopsorhamdia phantasia*). The Río Tahuamanu above Filadelfia contains a number of rocky outcrops, including some small rapids and shallow regions with many downed logs. In the dry season the area near Aserradero was almost un navigable because of the tree trunks and rapids. Such regions are prime candidates for spawning areas of the commercially important, large, pimelodid catfishes (e.g., *Pseudoplatystoma fasciatum, Leiarius marmoratus*) (Barthem and Goulding, 1997). As development continues towards the Peruvian border, pressure may be exerted to channelize the river to permit shipment of supplies or equipment up river.

CONCLUSIONS AND RECOMMENDATIONS

The following are the major conclusions and recommendations that derive from our studies:

1. The region of the Río Tahuamanu-Río Manuripi, Upper Río Orthon basin, Pando, Bolivia is a potential hotspot for the biodiversity of freshwater fishes. We discovered 313 species in the region, of which 87 species represent new records for Bolivia. This region contains 63% of the fishes found in the Bolivian Amazon and 49% of the species found in the entire country.

2. The diversity within the Río Tahuamanu-Río Manuripi region will increase with increased sampling. The collecting efforts at the end of the expedition were still increasing the species known from the area at a rate of 10.5 species per day. An asymptote was not reached.

3. The ichthyofauna is a unique assemblage of species. It appears to contain three distinct biogeographical elements as follows: i) widespread lowland Amazonian elements from the north and east; ii) Brazilian Shield elements from the Río Guapore/Itenez; and iii) headwater elements.

4. The freshwater fishes are economically valuable. Many of the species are currently used for subsistence or commercial fishery and have a high value. The food fishery may largely be for exportation. The ornamental fishes have a very high value.

5. Populations and stocks of commercially exploited fishes need to be studied immediately across the Río Madeira basin in Bolivia and coordinated with Brazil. The exploitation of local fishes for exportation may be putting undo pressure on stocks. Catch data from the expedition do not support viable populations living in the region.

6. We recommend the development of fisheries for local consumption and the careful regulation of catches for exportation.

7. The potential to develop local fishery for ornamental species should be studied. Many species are available locally that are not well - represented in the aquarium trade, but population biology and life histories need to be studied to guarantee sustainability.

8. Zones of critical habitats with varzea, cochas, main channels and upland areas should be protected. Rather than the formation of a park, consider multiple use zones with some habitats restricted for modification. Protect the narrow floodplain.

9. Programs should be developed to work with local fisherman and floodplain residents to explain the relationships between maintenance of habitats and biodiversity of fishes. Educational programs should also be developed and should involve local residents and fishermen in monitoring stocks and habitats. These programs should promote the learning of the different fishes and how to recognize new or unusual species that should be brought to the attention of scientists.

LITERATURE CITED

Barthem, R., and M. Goulding. 1997. The catfish connection: ecology, migration, and conservation of Amazon predators. Biology and Resource Management in the Tropics Series, Columbia University Press, New York.

Eigenmann, C. H., and G. S. Myers. 1929. The American Characidae. Memoirs of the Museum of Comparative Zoology, Harvard, 43: 429-574.

Gery, J. 1977. Characoids of the world. T.F.H. Publications, New Jersey. 672 pp.

Goulding, M. 1980. The fishes and the forests: explorations in Amazonian natural history. University of California Press, Los Angeles. 280 pp.

Goulding, M. 1981. Man and fisheries on an Amazon frontier. Dr. W Junk Pub., Boston. 137 pp.

Goulding, M., M. L. Carvalho, and E. G. Ferreira. 1988. Río Negro: rich life in poor water: Amazonian diversity and foodchain ecology as seen through fish communities. SPB Academic Publishing, The Hague, Netherlands. 200 pp.

Lauzanne, L., G. Loubens, and B. Le Guennec. 1991. Liste commentée des poissons de l'Amazonie bolivienne. Revista Hydrobiologia Tropical 24:61-76.

López-Rojas, H., J. G. Lundberg, and E. Marsh. 1984. Design and operation of a small trawling apparatus for use with dugout canoes. North American Journal of Fisheries Management 4: 331-334.

Lowe-McConnell, R. H. 1987. Ecological studies in tropical fish communities. Cambridge University Press, New York. 382 pp.

Machado-Allison, A., J. Sarmiento, P.W. Willink, N. Menezes, H. Ortega, and S. Barrera. 1999. Diversity and abundance of fishes and habitats in the Río Tahuamanu and Río Manuripi basins. *In* Chernoff, B., and P.W. Willink (eds.). A biological assessment of of the Upper Río Orthon basin, Pando, Bolivia. Pp. 47-50. Bulletin of Biological Assessment No. 15, Conservation International, Washington, D.C.

Mago-Leccia, F. 1994. Electric fishes of the continental waters of America. Biblioteca de la Academia de Ciéncias Físicas, Matematicas y Naturales, Volume XXIX, Caracas, Venezuela.

Peres, C. A., and J. W. Terborgh. 1995. Amazonian nature reserves: an analysis of the defensibility status of existing conservation units and design criteria for the future. Conservation Biology 9: 34-45.

Santos, G. M., M. Jegu, and B. Merona. 1984. Catálogo de peixes comerciales do baixo Tocantins. Elcectronorte/ CNPq/INPA, Manaus. 83 pp.

Sarmiento, J. 1998. Ichthyology of Parque Nacional Noel Kempff Mercado., Appendix 5. *In* Killeen, T.J. and T. S. Schulenberg (eds.) A biological assessment of Parque Nacional Noel Kempff Mercado. Pp. 168-180. RAP Working Papers No. 10. Conservation International, Washington, D.C.

Stewart, D. J. 1985. A new species of *Cetopsorhamdia* (Pisces: Pimelodidae) from the Río Napo Basin of Eastern Ecuador. Copeia 1985: 339-344.

Stewart, D. J., R. Barriga, and M. Ibarra. 1987. Ictiofauna de la cuenca del Río Napo, Ecuador Oriental: lista anotada de especies. Revista del Politecnica 12: 9-63.

Vari, R. P. 1992. Systematics of the Neotropical characiform genus *Curimatella* Eigenmann and Eigenmann (Pisces: Ostariophysi) with summary comments on the Curimatidae. Smithsonian Contributions to Zoology 533: 1-48.

Walters, P. R., R. G. Poulter, and R. R. Coutts. 1982. Desarrollo pesqueiro de la Region Amazonica en Bolivia. London: Tropical Products Institute, Overseas Development Agency.

Welcomme, R.L. 1990. Status of fisheries in South American rivers. Interciencia 15: 337-345.

CHAPTER 4

DIVERSITY AND ABUNDANCE OF FISHES AND HABITATS IN THE RIO TAHUAMANU AND RIO MANURIPI BASINS

Antonio Machado-Allison, Jaime Sarmiento, Philip W. Willink, Naércio Menezes, Hernán Ortega, and Soraya Barrera

ABSTRACT

Fishes were collected at 85 stations in the Río Tahuamanu and Río Manuripi basins. These basins were divided into five subregions. The physical features, number of species, number of specimens, and taxa present for each station are used as a basis for a brief description of each region's fish community.

INTRODUCTION

Here we present a summary description of the five subregions in the Río Tahuamanu and Río Manuripi basins. The predominant aquatic physical features are elaborated upon to provide context to the biological information. Number of species and abundance in specified habitats will be used along with comments on particular taxa to synthesize a brief description of each region's ichthyological community. These findings are used as a basis for conservation recommendations.

Fishes were collected at 85 stations in the Río Tahuamanu and Río Manuripi basins. Each station is described in detail in Appendix 8. Latitudes and longitudes are not available for some of the stations due to interference between the GPS units and their appropriate satellites. The number of localities exhibiting particular macrohabitat types and water characteristics for the entire Upper Río Orthon basin is given in Table 4.1.

Table 4.1. Number of localities exhibiting the indicated macrohabitats and water characteristics in the entire Upper Río Orthon basin.

Macro-habitat	Water characteristics		
	Black Water	White Water	Turbid Water
River	6	37	32
Lake & Cocha	2	6	4
Tributary	6	13	12
Flooded Lake	12	6	1
Flooded Forest	1	1	1

Upper Nareuda (13 Sampling Stations, P02-01 to P02-13)

The region includes small creeks and rivers (1.5 to 8 m wide). Most have whitewater and turbid conditions, but there is a caño and a blackwater igarape (Table 4.2). Sandy/muddy shores and bottoms are common. Presence of grasses and aquatic plants is rare. Some riparian forest is present, particularly along small creeks flowing out of the forests. Water current is dependent upon the area sampled, ranging from swift in the main channel to almost stagnant in caños.

The number of species and specimens collected at each station can be found in Appendix 8. The number of taxa in small, blackwater rivers ranges from 19 to 33 species (mean = 24) and 43 to 425 specimens (mean = 131.4). Caños range from 8 to 18 species (mean = 13) and 15 to 119 specimens (mean = 52.8). In the igarape preto, 26 species and 93 specimens were collected.

The number of species of fishes is low in the small, blackwater rivers. However, there are several species of economic importance, such as *Knodus gamma, Odontostilbe hasemani, Gasteropelecus sternicla, Aphanotorulus frankei, Prionobrama filigera, Rineloricaria lanceolata,* and *Otocinclus mariae*. Several species of *Apistogramma* and *Aequidens* are very common and always abundant. Caños, on the other hand, possess a very low diversity and productivity.

Table 4.2. Number of localities exhibiting the indicated macrohabitats and water characteristics in the Nareuda sub-basin.

Macro-habitat	Water characteristics		
	Black Water	White Water	Turbid Water
River	-	8	8
Lake & Cocha	-	2	1
Tributary	1	12	12
Flooded Lake	-	1	-
Flooded Forest	-	-	-

Lower Nareuda (11 Sampling Stations, P01-11 to P01-21)

The region includes a mixture of medium-sized rivers (12-15 mts wide), creeks, rapids, dead arms, and lagoons (Table 4.2). Sandy/muddy shores and bottoms are common, as well as riparian forest. At some stations, such as creeks flowing out of forests and lagoons, logs and leaves are abundant along the shore. This debris provides important microhabitats for several species of silurids, characoids, and cichlids. Water current is dependent upon the area sampled, ranging from fast in the main channel to stagnant in lagoons and dead arms.

Note: One station (P12) was classified as a whitewater curiche. This is questionable because the field notes contradict themselves by stating that this area is formed from water from the Río Nareuda, which is a blackwater river.

The number of species and specimens collected at each station can be found in Appendix 8. The single creek sampled had 19 species and 34 specimens. The one rapids sampled had 19 species and 71 specimens. Medium-sized rivers range from 9 to 38 species (mean = 24.3) and 21 to 147 specimens (mean = 77.4). Lagoons range from 34 to 43 species (mean = 38) and 279 to 444 specimens (mean = 381). Lagoons and dead arms appear to be the most biodiverse and productive areas.

Several groups of species are common and abundant in the rivers and creeks. Examples are *Astyanax abramis, Odontostilbe paraguayensis, Phenacogaster* spp., *Cyphocharax* spp., *Hypoptopoma joberti, Tyttocharax madeirae,* and *Corydoras* spp. The number of important aquarium species increases in lagoons or dead arms. In addition to the species present in the creeks and rivers, there are other groups, such as cichlids and electric fishes, which increase the diversity and the importance of these flooded areas. Some of the additional species are *Apistogramma* spp., *Aequidens* spp., *Moenkhausia* spp., Gasteropelecins, *Mesonauta festivus, Agamyxis* sp., *Hypostomus* sp., *Hoplosternum* spp., *Liposarcus disjunctivus, Peckoltia arenaria, Tatia perugiae, Auchenipterus nuchalis, Parotocinclus* sp., and *Rineloricaria* spp. The abundance of several species of *Corydoras* together with rare species of *Peckoltia, Hypoptopoma,* and *Otocinclus* is significant in light of the popularity of these species in the aquarium trade.

Upper Tahuamanu (10 Sampling Stations, P01-01 to P01-10)

The region includes large rivers (70+ m wide), creeks, dead arms, and lagoons (Table 4.3). Sandy/muddy shores and bottoms are common, as well as riparian forest. At some stations, such as creeks flowing out of forests, logs and leaves are abundant along the shore. This debris provides important microhabitats for several species of siluroids, characoids, and cichlids. Water current is dependent upon the area sampled, ranging from rapid in the main channel to stagnant in the lagoons and dead arms. White water predominates (Table 4.3).

The number of species and specimens collected at each station can be found in Appendix 8. The number of taxa collected ranges in creeks from 5 to 30 (mean = 20.3), in large rivers from 14 to 36 (mean = 24.3), and in lagoons from 25 to 32 (mean 28.5). Specimen abundance showed the following results: creeks, 66 to 136 specimens (mean = 108.7); large rivers, 25 to 217 specimens (mean = 115.3); lagoons, 85 to 389 specimens (mean = 237).

Several groups of species are very common and abundant in the rivers and creeks. Examples are *Astyanax abramis, Odontostilbe* spp., *Prionobrama filigera, Steindachnerina* spp., and *Pimelodella* spp. The number of species important to the aquarium trade and utilized for human consumption increases in the lagoons or dead arms. In addition to the species present in the creeks and rivers, there are other groups, such as cichlids and electric fishes, which increase the diversity and the importance of these flooded areas. These additional taxa include *Apistogramma* spp., *Aequidens* spp., *Moenkhausia* spp., *Mesonauta festivus, Agamyxis* sp., *Hypostomus* sp., *Hoplosternum* spp., *Liposarcus disjunctivus, Potamorhina* spp., *Plagioscion squamosissimus, Auchenipterus nuchalis, Prochilodus nigricans, Pygocentrus nattereri,* and *Hydrolycus* spp.

Table 4.3. Number of localities exhibiting the indicated macrohabitats and water characteristics in the Upper Tahuamanu sub-basin.

Macro-habitat	Water characteristics		
	Black Water	White Water	Turbid Water
River	-	5	2
Lake & Cocha	-	1	-
Tributary	2	1	-
Flooded Lake	-	3	-
Flooded Forest	-	-	-

Middle Tahuamanu (14 Sampling Stations, P02-14 to P02-27)

The region includes small creeks, rapids, and large rivers (up to 100 mts wide). White, turbid water is most common, but a blackwater igarape was also surveyed (Table 4.4). Sandy/muddy shores and bottoms are common. Presence of grasses and aquatic plants is very rare. Some riparian forest is present, particularly along small creeks flowing out of the forests. Water current is dependent upon the area sampled, ranging from very fast in the rapids to medium in the main channel.

The number of species and specimens collected at each station can be found in Appendix 8. The number of taxa in large rivers ranges from 7 to 29 species (mean = 19.4) and 21 to 379 specimens (mean = 155.1). In the rapids, 31 species and 185 specimens were collected. The number of taxa in the blackwater igarape ranges from 16 to 21 species (mean = 18.5) and 74 to 84 specimens (mean = 79). The single lake sampled possessed a very low diversity and productivity (20 species and 90 specimens). This area was heavily damaged by logging and cattle ranching.

The most abundant species are *Pimelodella itapicuruensis, Acanthopoma bondi, Pimelodella gracilis, Odontostilbe hasemani, Aphanotorulus frankei,* and a new species of *Megalonema* (one station had 36 specimens of this species).

Table 4.4. Number of localities exhibiting the indicated macrohabitats and water characteristics in the Middle Tahuamanu sub-basin.

Macro-habitat	Water characteristics		
	Black Water	White Water	Turbid Water
River	-	10	10
Lake & Cocha	-	1	1
Tributary	3	-	-
Flooded Lake	-	1	1
Flooded Forest	-	-	-

Manuripi (including Lower Tahuamanu) (37 Sampling Stations, P01-22 to P01-39 and P02-28 to P02-46)

The region includes large rivers (50 to 75+ m wide), dead arms, and lagoons (Table 4.5). Sandy/muddy shores and bottoms are abundant. Grasses, aquatic plants, and riparian forests are common near Puerto Rico, but rare elsewhere.

Grasses and aquatic plants are most abundant in lagoons and backwaters. Riparian forest is most common along small creeks coming out from the forest. At some stations, such as lagoons or backwaters, logs and leaves can be found along the shore.

As pointed out before, this debris provides important microhabitats for several species of siluroids, characoids, and cichlids. Water is often black, although the Río Orthon and Río Tahuamanu are whitewater rivers (Table 4.5). Water current is dependent upon the area sampled, ranging from fast in the main channel to stagnant in lagoons and dead arms.

The number of species and specimens collected at each station can be found in Appendix 8. For stations P1-22 to P1-39, the number of taxa in blackwater rivers ranges from 25 to 60 species (mean = 44) and 61 to 834 specimens (mean = 451). Lagoons range from 23 to 45 species (mean = 36.4) and 357 to 1,014 specimens (mean = 881.7). In the Río Orthon, 62 species and 332 specimens were collected, whereas 16 species and 40 specimens were collected in the Río Tahuamanu.

For stations P2-28 to P2-46, the number of taxa in large rivers ranges from 18 to 38 species (mean = 30) and 75 to 551 specimens (mean = 235). Lagoons range from 21 to 43 species (mean = 32.4) and 232 to 1083 specimens (mean = 572.2).

Blackwater rivers had the greatest species richness, including several species of economic importance. Examples of species collected in the blackwaters are *Corydoras loretoensis, Brachyrhamdia marthae, Hemigrammus unilineatus, Amblydoras hancockii, Pyrrhulina vittata, Moenkhausia colletti, M. sanctaefilomenae, Hemigrammus ocellifer, Acanthodoras cataphractus, Carnegiella myersi, Mesonauta festivus, Hypoptopoma joberti, Prionobrama filigera, Rineloricaria lanceolata, Entomocorus benjamini, Apteronotus albifrons, Eigenmannia virescens, Nannostomus trifasciatus,* and *Crenicara unctulata.* Several species of *Apistogramma* and *Aequidens* are also very common and always abundant. A number of important aquarium species were collected in lagoons or dead arms, and the specimens tended to be large.

The area near Puerto Rico has been moderately damaged by cattle ranching. However, species richness is still high in some habitats (e.g. lagoons (cochas)). The most abundant species are *Moenkhausia colletti, Moenkhausia lepidura, Apistogramma* spp., *Carnegiella myersi, Doras* cf. *carinatus, Opsodoras stubelii, Eigenmannia* spp., *Entomocorus benjamini, Pimelodella gracilis, Poptella compressa, Prionobrama filigera, Knodus victoriae, Tympanopleura* sp., and *Rineloricaria* spp. Number of specimens is highest in cochas and lagoons. Electric fishes are also common here.

Table 4.5. Number of localities exhibiting the indicated macrohabitats and water characteristics in the Manuripi sub-basin.

Macro-habitat	Water characteristics		
	Black Water	**White Water**	**Turbid Water**
River	**6**	**12**	**12**
Lake & Cocha	**2**	**2**	**2**
Tributary	-	-	-
Flooded Lake	**12**	**1**	-
Flooded Forest	1	1	1

CONSERVATION RECOMMENDATIONS

A general recommendation for all the sub-basins is to maintain the hydrological cycle responsible for the annual flooding which creates and maintains the lagoons and dead arms. These lagoons and dead arms serve as nursery and feeding areas for a large number of fishes. Many of these species are popular in the aquarium trade. An activity that could be promoted is the harvesting or aquaculture of ornamental species. This activity would best be conducted in the isolated lagoons, cochas, or dead arms of the river, and could be a source of income for the local people. Managed properly, this would also help to promote the conservation of the aquatic ecosystem.

Some species serve as a source of food for local people. Rivers provide water for the local inhabitants.

The Middle Tahuamanu has been severely damaged by cattle ranching. Restoration of the gallery forest and restriction of burning is highly recommended. Ashes may poison the waters.

Blackwater rivers and lagoons of the Manuripi are unique habitats and very fragile. Human development in the region has to be regulated. As in the Middle Tahuamanu, restoration of the gallery forest and restriction of burning is necessary.

CHAPTER 5

GEOGRAPHIC AND MACROHABITAT PARTITIONING OF FISHES IN THE TAHUAMANU-MANURIPI REGION, UPPER RIO ORTHON BASIN, BOLIVIA

Barry Chernoff, Philip W. Willink, Jaime Sarmiento, Antonio Machado-Allison, Naércio Menezes, and Hernán Ortega

ABSTRACT

The distribution of fishes in the Upper Río Orthon basin has important ramifications for regional conservation recommendations. Simpson's index of similarity and a measure of matrix disorder (or entropy) were used to test if fishes were homogeneously distributed among geographic subregions, macrohabitats, or among classes of water. Species were distributed non-randomly in regards to subregion, with the most species found in the Río Manuripi. It is believed to be a hierarchical pattern with the faunal similarities nested within the larger fauna represented by the Río Manuripi. Species were also distributed non-randomly in relation to macrohabitat, with the highest species richness observed in rivers. We were unable to determine unambiguously if this pattern was clinal or nested because there was no obvious way to order the relationships of habitats, but the data are most consistent with fauna being derived from riverine habitats on seasonal flooding cycles. Species were distributed homogeneously among black, white, and turbid waters. This finding was unexpected. Based on the distribution of fishes, it is recommended that the Río Manuripi and Río Nareuda be designated as core conservation areas because combined they represent 75% of the regional diversity. Tributaries are the most endangered macrohabitat because of their fragility, uniqueness, and the current trends in habitat destruction. More than 80% of the fishes are dependent upon the flooded areas for reproduction, nursery grounds, or critical foraging, which highlights the importance of seasonal inundations on the regional aquatic ecology.

INTRODUCTION

When documenting the overall diversity of the freshwater fishes found in the Ríos Tahuamanu, Nareuda and Manuripi, Chernoff et al. (1999) associated the high species richness (313 species) in a general way with broad zoogeographic distributions and with habitat heterogeneity. In that paper and in Machado-Allison et al. (1999), the areas surveyed during the September 1996 AquaRAP were subdivided into six regions: i) the Upper Río Tahuamanu; ii) the Middle Río Tahuamanu; iii) the Lower Río Tahuamanu; iv) the Upper Río Nareuda; v) the Lower Río Nareuda; and vi) the Río Manuripi (Maps 1-3). Intra-regional differences in the ichthyofauna were established (Appendix 6). These papers also documented the principal threats to the fauna as logging, ranching, burning, and overfishing.

A pattern of heterogeneous distribution by the fishes within the Tahuamanu-Manuripi region would have important ramifications for conservation recommendations. For example, if the fauna were homogeneously distributed, then a core conservation area could be established that might effectively protect the vast majority of the species. However, as the distribution of the species either among regions or among habitats becomes increasingly distinct and patchy, then a single core area, apart from the entire region, may not provide the desired level of protection. Can we use information on the relative heterogeneity of distributions among sub-regions or among macrohabitats to predict possible faunal changes in response to specific environmental threats? Can such analyses be carried out within the framework of a rapid assessment program?

This paper will test three null hypotheses that are critical to the conservation of freshwater fishes of the Tahuamanu-Manuripi region as follows: that the fishes are homogeneously distributed among (i) six subregions; ii) five macrohabitats; and (iii) three classes of water. To test these hypotheses we introduce two new approaches to the estimation of faunal similarity as well as use methods developed by Atmar and Patterson (1993). Because two of the three null hypotheses are rejected, we then estimate what the ichthyofauna might resemble given the effects of habitat destruction in the region.

METHODS

The basic data used to test these hypotheses were lists that correspond to that shown for the six subregions in Appendix 6. Although we counted all of the individuals captured at each field sampling station, we make no calculations of abundance or in any way use the number of times that a species was collected in a subregion, in a macrohabitat or in a type of water. We believe that for point source data obtained in rapid assessments that only the presence of a species should be used as information. We scored the presence of a species as "1" and its absence as "0" in matrices with species on the rows and subregion, macrohabitat or water-type on the columns.

A number of similarity indices for binary (presence/absence) data have been used in faunistic or floristic works (e.g., Goulding et al., 1988; Cox Fernandes, 1995). Only those indices that are based upon positive matches are appropriate for point-source data when reasonably complete estimates of the fauna do not exist. Incorporation of negative matches (shared absences) into analyses is not conservative because absence from an area may be because a species is not present or because a species was accidentally missed during sampling.

Of the association indices for binary characters using shared presences as information, the three most common are Jaccard's, Sorenson's, and Simpson's. They use the following table format:

		List 1	
		1	**0**
List 2	**1**	*a*	*b*
	0	*c*	*d*

where, *a* is the number of positive matches or species present in both lists, *b* is the number of species present in List 2 and absent from List 1, *c* is the converse of *b*, and *d* is the number of negative matches or species absent from both localities. The three indices are identical when the number of species present in each list (i.e., in state 1) is identical. If the numbers of presences differ between lists then the indices treat them in different ways.

Using the Jaccard index, similarity is $a/(a+b+c)$ (see Cox Fernandes, 1995). The shared presences are expressed as a percentage of the total number of species. Sorenson's index, $= 2a/(2a+b+c)$, attempts to increase the weight of shared species over the total fauna (see Goulding et al., 1988). Because the denominators weight the difference in faunal size, both Jaccard's and Sorenson's indices work best under the assumption that the compared faunas contain potentially the same number of species. Alternatively, let us compare the faunas from a river and a small, shallow, seasonally flooded lake or cocha (Table 5.1). All of the species from the cocha are found in the river but not conversely. Jaccard's and Sorenson's similarities are 0.5 and 0.67, respectively. But from the standpoint of the cocha, 100% of its fauna is found in and, in fact, is derived from the river. Thus, there may not be the environmental or ecological potential for two habitats, regions, etc. to accumulate the same number of species.

Table 5.1. Two hypothetical faunas found living in a river and a seasonally flooded lake or cocha.

	River	Cocha
species 1	1	0
species 2	1	1
species 3	1	0
species 4	1	1
species 5	1	0
species 6	1	1
species 7	1	1
species 8	1	0
species 9	1	1
species 10	1	0

In order to accommodate different accumulation potentials, we present a different approach. In the hydrological system of the Tahuamanu-Manuripi region all subregions, macrohabitats and water-types are connected to each other. Because we have point source data with repetitions over each of these three factors, it is conservative to assume an overall fauna from which subsets (e.g., homogeneous, random, nested, etc.) are drawn. We base our ideas about similarity from the smaller of two species lists because to do otherwise is to interpret the significances of the negatives as absent species, when actually the 0's in the matrices are coding artifacts or place holders for missing data. Simpson's index of similarity, S_s, provides the proper framework. $S_s = a/(a+b)$, where $b < c$, or $S_s = a/n$ where n is the number of species present in the smaller of two lists.

The interpretation of S_s is somewhat problematic and several questions arise. Does a very low S_s imply biological structuring? Does a very high S_s imply correlation or dependence? How do we interpret random effects? Can we place confidence limits on S_s? To answer these questions we use rarefaction and bootstrapping techniques.

If we compare the similarity of two lists, both of which are drawn from a fixed larger universe (e.g., the set of all species captured during the AquaRAP in the Tahuamanu-Manuripi region), then we calculate the similarity of the larger sample rarefied down to the smaller size. Returning to the trivial example of Table 5.1, we would draw n random, five-species-samples from the river habitat to compare with the cocha sample. We compare the observed similarity, S_s, to the distribution observed from the bootstrapped random samples. Using a 2-tailed parametric approach, the probability of obtaining the observed similarity at random is calculated from the number of standard deviations that the observed similarity was either above or below the mean of the bootstrap random distribution.

A random-similarity distribution was generated for each of the following sizes: 40, 75, 100, 116, 125, and 220. This range corresponds to the actual number of species observed in various subregions, macrohabitats, and water classes. Two hundred replications were used to produce each distribution (e.g., Fig. 5.1).

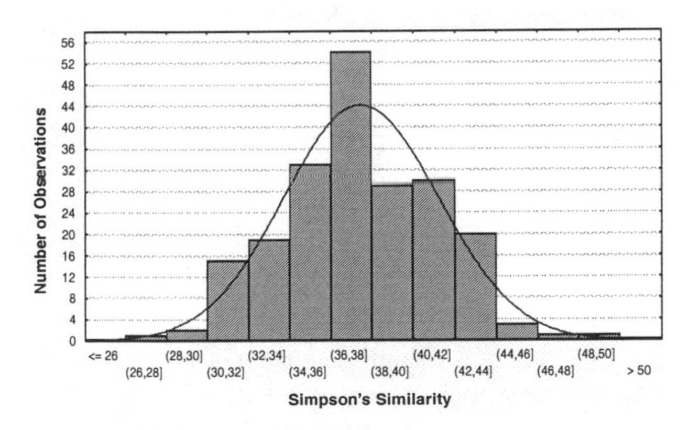

Figure 5.1. Distribution of Simpson's Similarity Index from 200 randomly simulated vectors of presence-absence data given the constraint that 116 species are present in each vector

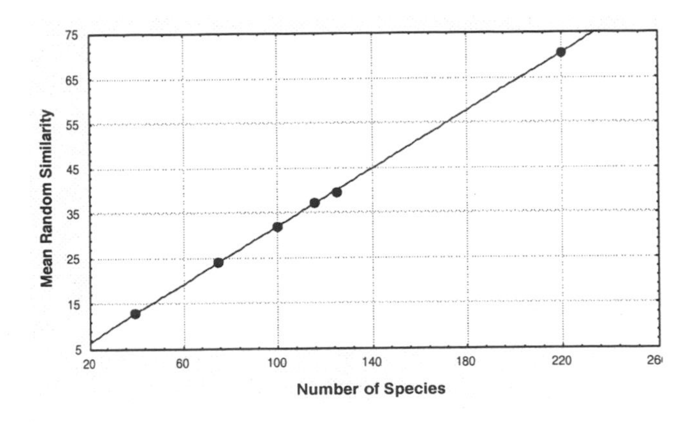

Figure 5.2. Mean random similarity, Simpson's Index, plotted against the number of species constrained to be present in the sample. Each mean was calculated from 200 random samples of a given size.

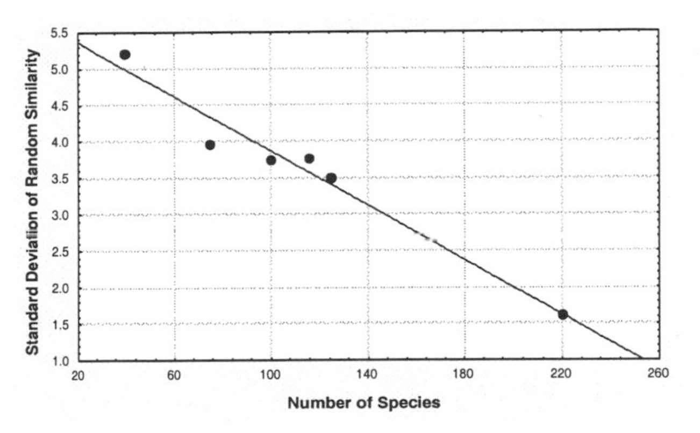

Figure 5.3. Standard deviation of random similarity, Simpson's Index, plotted against the number of species constrained to be present in the sample. Each standard deviation was calculated from 200 random samples of given size.

The means and standard deviations of these distributions were then plotted against number of species present in a sample (Figs. 5.2, 5.3). As the number of species present in a sample increases the observed similarity due to random effects also increases (Fig. 5.2), but the variance decreases (Fig. 5.3). Although the effects and the methods are for the general case, the slope of the curves and forms of the distributions (Figs. 5.1-5.3) are a function of universe size, i.e. 313 species.

To compare two faunal lists, we rarefied the larger sample down to the size of the smaller sample, calculated the mean S_s and compared that mean to the mean of random similarity for a fauna the size of the smaller list. If the rarefied mean is greater than or less than 2 standard deviations from mean random similarity the conclusion is made that the observed similarity is not due to random effects. If S_s falls within the random effects, then we fail to reject the null hypothesis that the two lists are equal, and we conclude that the two lists are drawn homogeneously from a larger distribution. If S_s exceeds the mean random similarity by 2 standard deviations or more, then we conclude that the similarity is due to biological dependence or correlation, such as nested subsets which comprised the example in Table 5.1.

That is one population forms the source population for another. If S_s is less than the random mean we reject the null hypothesis and can search for biological or environmental reasons for the dissimilarity.

If we discover that similarities are not random, we can investigate whether the pattern of species presences in relation to environmental variables is non-random. The measure of matrix disorder as proposed by Atmar and Patterson (1993) calculates the entropy of a matrix as measured by temperature. Temperature measures the deviation from complete order (0°) to complete disorder (100°) in which the cells of a matrix are analogous to the positions of gas molecules in a rectangular container. After the container has been maximally packed to fill the upper left corner (by convention), the distribution of empty and filled cells determines the degree of disorder in the species distributions and corresponds to a temperature that would produce the degree of disorder (Atmar and Patterson, 1993). To test whether the temperature could be obtained due to random effects, 500 Monte Carlo simulations of randomly determined matrices of the same geometry were calculated. The significance of observed temperatures is ascertained in relation to the variance of the simulated distributions. Two patterns are consistent with a rejection of the null hypothesis by this method: that the distributions are nested subsets (Atmar and Patterson, 1993), or there is a clinal turnover in fauna. Software to calculate matrix disorder is available from Atmar and Patterson at the following internet site: http://www.fieldmuseum.org.

RESULTS

I. Sub-Basin

The Tahuamanu-Manuripi region was divided into six sub-basins (Maps 2 - 3) as follows: Upper Río Tahuamanu, including the mouth of the Río Muymanu; Middle Río Tahuamanu; Lower Río Tahuamanu; Upper Río Nareuda, including small streams (garapes); Lower Río Tahuamanu; and Río Manuripi. With the exception of the Lower Río Tahuamanu, these subregions showed important patterns of similarities and differences that will be critical to take into consideration for the conservation of species of freshwater fishes in the region.

By far the most species rich sub-basin was the Río Manuripi where we collected 220 species of fishes (Table 5.2), 70.5% of all the species collected. With the exception of the Lower Río Tahuamanu that was highly degraded environmentally, the other sub-regions have almost the same number of species (range = 113-123; Table 5.2), each fauna approximately 45% the size of that of the Río Manuripi.

Table 5.2. Means of Simpson's index of similarity among ichthyofaunas found in six subregions of the Upper Orthon basin (lower triangle); larger samples were rarefied 200 times to the size of the smaller sample; index values reported as percentages. Numbers of species shared exclusively by pairs of sub-basins (upper triangle). Abbreviations: UT – Upper Río Tahuamanu; MT – Middle Río Tahuamanu; LT – Lower Río Tahuamanu; UN – Upper Río Nareuda; LN – Lower Río Nareuda; Man – Río Manuripi; n - number of species present; u – number of unique species; % - percentage of unique species.

	UT	MT	LT	UN	LN	Man
UT		2	0	8	3	8
MT	55.3		0	7	1	11
LT	72.1	74.4		0	0	3
UN	61.2	55.2	58.1		4	5
LN	60.2	49.6	51.2	55.8		15
Man	73.2	66.7	95.3	58.6	72.6	
n	123	123	43	116	113	220
u	6	19	0	15	10	78
%	4.9	15.5	0	12.9	8.9	35.5

The biggest factor that differentiates the Río Manuripi from the other areas concerns the number of unique species that were found within the regions (Tables 5.2, 5.3). Thirty-five percent of all the species that were found in the Río Manuripi were found there exclusively. Of these taxa, about 10 are highly artifactual as unique species, including the piranhas (*Serrasalmus* spp.), the tiger catfish (*Pseudoplatystoma fasciatum*), and some of the curimatids (e.g., *Curimatella, Psectrogaster*). Only the Upper Río Nareuda and the Middle Río Tahuamanu had unique species comprising more than 10% of their faunas (Tables 5.2, 5.4, 5.5). Thus, the number of species unique to the Río Manuripi is proportionally greater than that found in other sub-basins. Perhaps the most surprising finding is that given an exceptionally high number of species (n=313) found in a very limited geographic region, there was only a small group of species (n=33, Table 5.3) present in all the sub-basins. Though this list (Table 5.3) contains members of each of the ostariophysan orders, only a single species of cichlid (*Aequidens* sp. 1) is represented.

Table 5.3. Fish species found only in the Río Manuripi (n=78) and those found in all sub-basins (n=33).

Río Manuripi		All Sub-basins
Acanthodoras taphractus	*Hypopygus lepturus*	*Equidens* sp. 2
Adontosternarchus clarkae	*Hypostomus* sp. 4	*Astyanax* cf. *abramis*
Agamyxis pectinifrons	*Iguanodectes spilurus*	*Carnegiella myersi*
Ageneiosus cf. *caucanus*	*Laemolyta* sp.	*Cochliodon* cf. *cochliodon*
Amblydoras cf. *hancockii*	*Leporinus* cf. *fasciatus*	*Corydoras acutus*
Anadoras cf. *grypus*	*Mesonauta* cf. *insignis*	*Corydoras* cf. *loretoensis*
Ancistrus sp. 4	*Metynnis luna*	*Creagrutus* sp. 2
Apistogramma sp. 4	*Microgeophagus altispinosa*	*Cyphocharax spiluropsis*
Astrodoras asterifrons	*Moenkhausia* cf. *chrysargyrea*	*Eigenmannia virescens*
Astronotus crassipinnis	*Moenkhausia* cf. *colletti*	*Gasteropelecus sternicla*
Auchenipterichthys thoracatus	*Moenkhausia* cf. *comma*	*Hoplias malabaricus*
Brachyhypopomus brevirostris	*Moenkhausia* cf. *megalops*	*Hypostomus* sp. 2
Brachyhypopomus pinnicaudatus	*Moenkhausia* sp. 5	*Imparfinis stictonotus*
Bunocephalus coracoideus	*Moenkhausia* sp. 6	*Knodus* cf. *gamma*
Carnegiella strigata	*Moenkhausia* sp. 7	*Knodus* cf. *victoriae*
Cheirodon piaba	*Nannostomus trifasciatus*	*Loricaria* sp.
Cheirodon sp. 1	*Opsodoras* cf. *humeralis*	*Moenkhausia colletti*
Cheirodontinae sp.	*Trachelyopterus* cf. *galeatus*	*Moenkhausia dichroura*
Cichla cf. *monoculus*	*Pimelodella* cf. *boliviana*	*Moenkhausia sanctaefilomenae*
Cichlasoma severum	*Pimelodella* cf. *itapicuruensis*	*Moenkhausia* sp. 3
Corydoras hastatus	Pimelodidae sp.	*Ochmacanthus* cf. *alternus*
Corydoras cf. *napoensis*	*Pimelodus* sp. 1	*Odontostilbe paraguayensis*
Crenicara cf. *unctulata*	*Pimelodus* sp. 3	*Otocinclus mariae*
Curimatella alburna	*Platydoras costatus*	*Paragoniates alburnus*
Curimatella dorsalis	*Psectrogaster rutiloides*	*Phenacogaster* cf. *pectinatus*
Cynodon gibbus	*Pseudohemiodon* sp. 1	*Pimelodella gracilis*
Cyphocharax cf. *plumbeus*	*Pseudoplatystoma fasciatum*	*Pimelodella* cf. *itapicuruensis*
Dianema longibarbis	*Pyrrhulina australe*	*Prionobrama filigera*
Distocyclus conirostris	*Rhabdolichops caviceps*	*Rineloricaria lanceolata*
Doras eigenmanni	*Rivulus* sp.	*Rineloricaria* sp.
Eigenmannia humboldtii	*Satanoperca* cf. *acuticeps*	*Steindachnerina dobula*
Electrophorus electricus	*Scoloplax* cf. *dicra*	*Sturisoma nigrirostrum*
Glyptoperichthys lituratus	Serrasalminae sp.	*Triportheus angulatus*
Hemidoras microstomus	*Serrasalmus* cf. *marginatus*	
Hemigrammus cf. *pretoensis*	*Serrasalmus marginatus*	
Hemigrammus ? sp.	*Serrasalmus* sp.	
Hemigrammus cf. *unilineatus*	*Tatia aulopygia*	
Hyphessobrycon cf. *anisitsi*	*Trachydoras paraguayensis*	
Hyphessobrycon ? sp.	*Tridentopsis pearsoni*	

Table 5.4. Fish species found only in the middle (n=19) and upper (n=6) portions of the Río Tahuamanu.

Middle Río Tahuamanu	Upper Río Tahuamanu
Acanthopoma cf. *bondi*	*Astyanax* sp.
Aphanotorulus unicolor	*Auchenipterus* cf. *nuchalis*
Bunocephalus sp. 3	*Duopalatinus* cf. *malarmo*
Centromochlus cf. *heckelii*	*Plagioscion squamosissimus*
Cetopsorhamdia phantasia	*Pristobrycon* sp.
Crenicichla sp. 1	*Roeboides* cf. *myersi*
Hemigrammus cf. *megaceps*	
Hyphessobrycon cf. *gracilior*	
Hyphessobrycon cf. *tucunai*	
Moenkhausia sp. 4	
Panaque sp.	
Pimelodus altissimus (sp. nov.)	
Pimelodus sp. 2	
Pimelodus sp. 4	
Planiloricaria cryptodon	
Plectrochilus sp.	
Pseudohemiodon sp. 2	
Trachydoras cf. *atripes*	
Xiliphius cf. *melanopterus*	

Table 5.5. Fish species found living only in the upper (n=15) and lower (n=10) sections of the Río Nareuda.

Upper Río Nareuda	Lower Río Nareuda
Ancistrus sp. 1	*Aequidens* cf. *tetramerus*
Bryconamericus cf. *caucanus*	*Apistogramma* sp. 3
Bryconamericus cf. *pachacuti*	*Brachyglanis* ? sp.
Bunocephalus sp. 1	*Callichthys callichthys*
Chrysobrycon sp. 1	*Charax gibbosus*
Creagrutus sp. 1	*Cheirodon* sp. 2
Gymnotus cf. *anguillaris*	*Moenkhausia* sp. 1
Heptapterus longior	*Myleus* sp.
Heptapterus sp.	*Tatia* cf. *Perugiae*
Hyphessobrycon agulha	Tetragonopterinae sp. 1
Imparfinis sp.	
Loricariidae sp.	
Pimelodus armatus	
Pseudohemiodon sp. 3	
Tyttocharax sp. nov.	

The similarities among the faunas within these sub-basins are shown in Table 5.2. All of the similarities are significantly greater than that predicted by random association ($P < 0.01$, two-tailed), rejecting null hypotheses of similarity due to homogeneity. The closest similarity to that being random involves the Lower Río Nareuda and the Middle Río Tahuamanu (Table 5.2), two regions that are adjacent. The similarity between these sub-basins, 49.6% is still 3.3 standard deviations above the mean ($P < 0.01$, two-tailed).

The rejection of random association (homogeneity) and the higher than expected similarities lead to the conclusion that the faunas among regions are in some way dependent and may be nested. That is, the faunas are hierarchically related among regions perhaps due to size (see Atmar and Patterson, 1993) or habitat (see below). A good example is provided by the fauna of the Lower Río Tahuamanu. This region contains taxa that live in all regions (Table 5.3), in the Middle Tahuamanu or in the Manuripi. Only two species are found in the Lower Río Tahuamanu that are not found in the Río Manuripi. The Lower Tahuamanu shares three species exclusively with the Río Manuripi but not with any other sub-basin (Table 5.3). The Lower Río Tahuamanu does not contribute any unique taxa to the overall assemblage, whereas 25 species were uniquely found in the Middle and Upper Río Tahuamanu (Table 5.4). Thus, the Lower Río Tahuamanu is almost a perfect subset of the Río Manuripi.

Given that the similarities and, hence, differences among sub-basins are not random, we used entropy statistics to test if they form an overall pattern. If the distribution of species among all sub-basins, despite the level of similarity, is random, then there will be a high degree of disorder in the matrix and the temperature will tend towards 100 as entropy increases. The results of the entropy statistics show that the matrix had a temperature of 27.83. Remember that a perfectly ordered matrix has a temperature of 0. However, this temperature is significantly cooler (i.e., has more structure) than that predicted by 500 Monte Carlo simulations (Fig. 5.4) such that the probability of obtaining the observed temperature, 27.83, at random is 3.6×10^{-17}. Thus, the matrix of species distributions is patterned despite the idiosyncratic distributions of selected species (Tables 5.3-5.5).

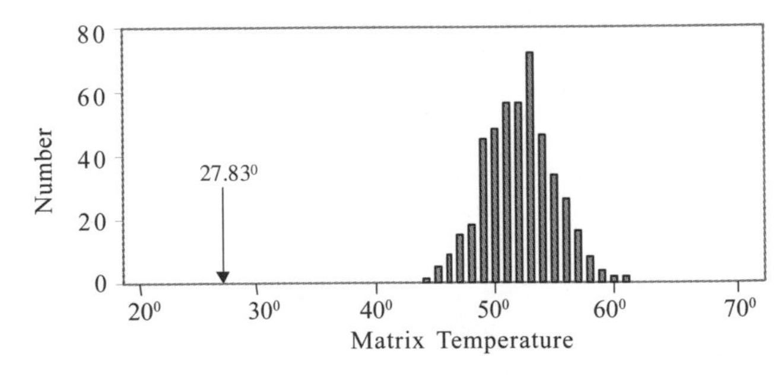

Figure 5.4. Distribution of matrix temperatures based upon 500 Monte Carlo simulations (histogram) displays a mean of 52.51^0 and a standard deviation of 2.95^0. The observed matrix temperature, 27.83^0, is 8.36 standard deviations below the random mean. The probability of obtaining the observed temperature by chance alone is 3.55×10^{-17}.

Two possibilities inhere as to the distributional pattern: nested subsets as noted by Atmar and Patterson (1993) or clinal variation (i.e., smooth turnover). Clinal transitions in fauna should roughly correspond to an isolation by distance model (Sokal and Oden, 1976). This pattern is consistent with high similarity among adjacent or proximate regions due to sharing of species. The similarity declines smoothly as distance among regions increases; there may be little or no species common to distant endpoints. A variety of cluster analyses from the distance matrix ($1-S_s$) produce identical cluster patterns that would superficially seem to lend support for clinal transition (Fig. 5.5) – a close relationship of the Río Manuripi with the Río Tahuamanu, the middle section intermediate to the Río Nareuda, Lower then Upper, which approximates the confluence of these rivers (Fig. 5.5).

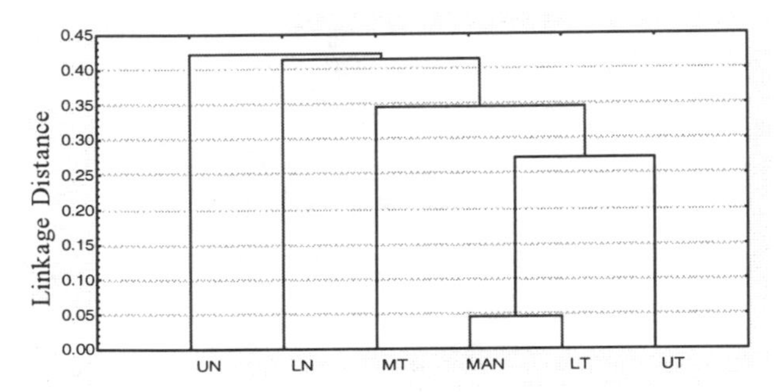

Figure 5.5. Cluster diagram, UPGMA, for dissimilarities among the six sub-basins: UN - upper Río Nareuda; LN - lower Río Nareuda; UT - upper Río Tahuamanu; MT - middle Río Tahuamanu; LT - lower Río Tahumanu; MAN Río Manuripi.

However consistent that pattern is with geography, clinal variation can be rejected from several sources of evidence. The Río Manuripi has the highest average similarities overall indicating that species are most often shared between each of the sub-basins and the Río Manuripi (Table 5.2). The Río Manuripi also shares more species exclusively with each of the other sub-basins, with the exception of the Upper Río Nareuda that shares almost as many species with the Upper and Middle Río Tahuamanu (Table 5.2).

When the number of species shared between and among regions, not including the 33 species living everywhere and not including the Lower Río Tahuamanu because of its small size, is plotted on a map of the area (Maps 1-3), we see that the pattern of similarity as a function of adjacency does not hold. For example, only one of the three largest sets of shared species is from adjacent areas (Fig. 5.6a). The largest set, other than the Lower Río Tahuamanu, belongs to the Upper Río Tahuamanu and the Río Manuripi. Even though this set contains 57 species, and the Upper Río Tahuamanu shares 35 species with the Middle Río Tahuamanu, which in turn shares 49 species with the Río Manuripi, only 24 species extend across these three regions (Fig. 5.6b). A more extreme circumstance inheres from the Middle Río Tahuamanu to the Upper Río Nareuda, a stretch of the basin only occupied by 12 shared species despite overall large faunal sizes (≥ 113) and relatively large shared sets by adjacent regions (Fig. 5.6b). These examples show that faunal turnover extended beyond two adjacent regions is not smooth. In fact, the most distant sites share the largest or among the largest sets of species.

This analysis leads to the conclusion that the pattern of species distributions is, nonetheless, structured across the sub-basins. The pattern is a hierarchical pattern with the faunal similarities nested within the larger fauna represented by the Río Manuripi. This conclusion is completely consistent with the cluster analysis (Fig. 5.5) because the order of sub-basins joining the cluster is exactly ordered by the sizes of the sets of shared species with the Río Manuripi. This nested pattern is significant from that due to random expectation but is coupled with idiosyncratic distributions that will be further resolved in the analysis of macrohabitats.

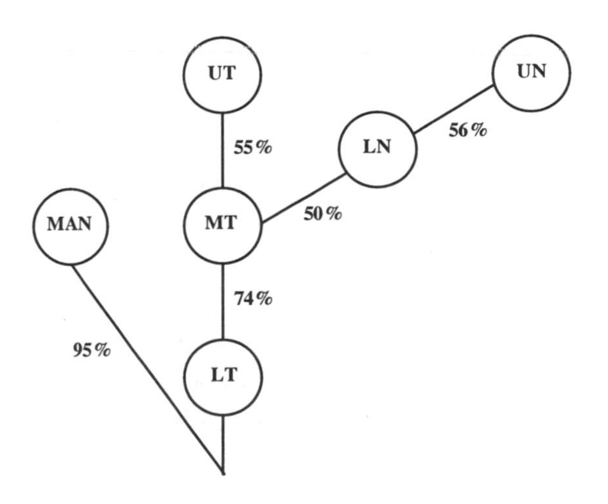

Figure 5.6 a. The network among the sub-basins abbreviated within the circles represents the straight-line river connections among the sub-basins. The values along the connections represent percent similarities, Simpson's index, between the connected sub-basins, including widely distributed species. Abbreviations: MAN - Río Manuripi; UT - Upper Río Tahuamanu; MT - Middle Río Tahuamanu; LT - Lower Río Tahuamanu; UN - Upper Río Nareuda; LN - Lower Río Nareuda.

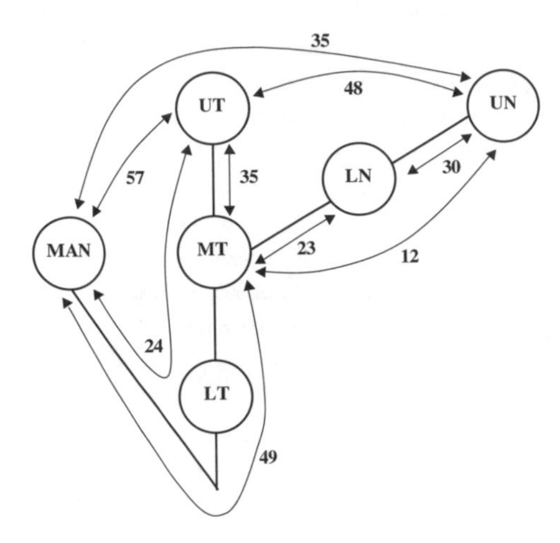

Figure 5.6 b. The numbers of species shared by pairs of sub-basins, connected by the arrows, not including the 33 species common to all sub-basins. Abbreviations: MAN - Río Manuripi; UT - Upper Río Tahuamanu; MT - Middle Río Thuamanu; LT - Lower Río Tahuamanu; UN - Upper Río Nareuda; LN - Lower Río Nareuda.

II. Macrohabitats

We recognized five broad categories of macrohabitats as follows: river, cocha, lake, tributary, and flooded forest. Cochas are oxbow lakes that were former river channels that became isolated during the meander history of the main river. The cochas all communicate with their main river during high water periods but were variable in their connections during low water periods. The openings of the cochas to the rivers were either shallow channels or were disconnected. Lakes were taken as different habitats from cochas in that lakes were not obviously oxbows, appeared to have a different formation, and were largely endorheic during low water. Lakes and cochas differed from flooded forest in that the latter were broadly open to the rivers and were simply flooded areas within the forest or within large forested islands. Flooded forest areas are only seasonally inundated.

Table 5.6. Means of Simpson's index of similarity among ichthyofaunas found in five macrohabitats of the Upper Orthon basin (lower triangle); larger samples were rarefied 200 times to the size of the smaller sample; index values reported as percentages. Numbers of species shared exclusively by pairs of macrohabitats (upper triangle). Abbreviations: R – river; C – cocha; T – tributary; L – lake; F – flooded forest; n – number of species; u – number of unique species; % — percentage of unique species.

	R	C	T	L	F
R		22	34	2	1
C	82.1		2	1	2
T	75.0	53.8		0	0
L	64.6	80.2	61.7		0
F	90.0	80.0	65.0	90.0	
n	223	106	128	178	40
u	91	13	29	1	2
%	40.8	12.3	22.7	0.6	5.0

Rivers were the most diverse macrohabitat followed by the tributaries; the flooded forests were the most depauperate owing to the fact that we were collecting during the driest period (Table 5.6). Rivers contained 71.5% of the entire fauna and 29% of the 91 unique taxa (Table 5.7). The percentage of unique species will drop with continual sampling. For example, *Rivulus* and *Thoracocharax* are not typical riverine species. However, many of the species are classical river species such as, *Adontosternarchus, Anchoviella, Centromochlus, Clupeacharax, Duopalatinus, Myleus, Steindachnerina,* and *Xiliphius* (Table 5.7). A number of the riverine taxa were captured in trawl samples. Surprisingly, there were a large number of ornamental species in the rivers (e.g., *Moenkhausia* and *Hyphessobrycon*), which may reflect degradation or elimination of their prime habitats.

Table 5.7. Fish species found living only in riverine macrohabitats (n=91).

Acanthopoma cf. *bondi*	*Moenkhausia* sp. 2
Adontosternarchus clarkae	*Moenkhausia* sp. 4
Ageneiosus cf. *caucanus*	*Moenkhausia* sp. 7
Anchoviella cf. *carrikeri*	*Moenkhausia* sp. 8
Aphanotorulus unicolor	*Myleus* sp.
Aphyocharax pusillus	*Opsodoras* cf. *humeralis*
Apteronotus bonapartii	*Opsodoras* cf. *stubelii*
Auchenipterichthys thoracatus	*Pachyurus* sp.
Brachyglanis ? sp.	*Panaque* sp.
Brachyhypopomus brevirostris	*Peckoltia arenaria*
Brachyhypopomus pinnicaudatus	*Phenacogaster* sp. 3
Brachyhypopomus sp.	*Pimelodella* cf. *gracilis*
Brochis splendens	*Pimelodella hasemani*
Bryconamericus cf. *pachacuti*	*Pimelodella* cf. *itapicuruensis*
Bunocephalus sp. 2	*Pimelodella* cf. *serrata*
Callichthys callichthys	Pimelodidae sp.
Carnegiella strigata	*Pimelodus "altipinnis"*
Centromochlus cf. *heckelii*	*Pimelodus altissimus* (sp. nov.)
Cetopsorhamdia phantasia	*Pimelodus* cf. *pantherinus*
Cheiroceros eques	*Pimelodus* sp. 1
Cheirodon sp. 1	*Pimelodus* sp. 2
Clupeacharax anchoveoides	*Pimelodus* sp. 3
Crenicichla sp. 1	*Pimelodus* sp. 4
Crossoloricaria sp.	*Planiloricaria cryptodon*
Curimatella immaculata	*Plectrochilus* sp.
Distocyclus conirostris	*Potamotrygon motoro*
Doras eigenmanni	*Psectrogaster rutiloides*
Duopalatinus cf. *malarmo*	*Pseudohemiodon* cf. *lamina*
Engraulisoma taeniatum	*Pseudohemiodon* sp. 1
Entomocorus benjamini	*Pseudohemiodon* sp. 2
Eucynopotamus biserialis	*Pseudostegophilu nemurus*
Farlowella sp. 2	*Pyrrhulina australe*
Hemidoras microstomus	*Rhabdolichops caviceps*
Hyphessobrycon cf. *gracilior*	*Rivulus* sp.
Hyphessobrycon ? sp.	*Roeboides* sp. 1
Hypopygus lepturus	*Roeboides* sp. 3
Hypostomus sp. 3	*Steindachnerina leucisca*
Laemolyta sp.	*Tatia* cf. *perugiae*
Lamontichthys filamentosus	*Thoracocharax stellatus*
Leiarius marmoratus	*Trachelyopterus* cf. *galeatus*
Leporinus cf. *fasciatus*	*Trachydoras* cf. *atripes*
Leporinus cf. *nattereri*	*Trachydoras paraguayensis*
Megalonema sp. nov.	*Tympanopleura* sp.
Moenkhausia cf. *colletti*	*Vandellia cirrhosa*
Moenkhausia cf. *jamesi*	*Xiliphius* cf. *melanopterus*
Moenkhausia cf. *megalops*	

Tributaries were largely restricted to the Río Nareuda system and the Upper Río Tahuamanu. The Río Nareuda is a blackwater system with many smaller tributaries and streams (garapes). Both the Nareuda and the garapes cross higher ground in the upper part of the drainage, crossing slightly dryer forest before descending into the floodplain of the Río Tahuamanu.

The network of tributaries contained 178 species overall, of which 29 (Table 5.8) were unique to that macrohabitat. Many of the unique species are smaller and diminutive and possibly valuable as ornamental species, including the four species of glandulocaudines (*Chrysobrycon* and *Tyttocharax*), the *Microschemobrycon, Characidium*, and the *Imparfinis*. There were also a few species that might be artifactually unique to the habitat, including *Cynopotamus gouldingi, Gymnotus* cf. *coatesi*, and *Pimelodus armatus*.

Taken together the cochas and lakes had 199 species, a total that comprises 64% of the fauna. There were 178 species in the lakes and 106 species in the cochas (Table 5.6). The majority of the species of cichlids were found in these habitats (Appendix 8; Machado-Allison et al., 1999) as well as a number of valuable miniatures and ornamentals, such as *Scoloplax, Corydoras,* and *Bunocephalus*. Despite the great wealth of diversity in these quiet-water, often weedy habitats, only a small proportion of it was found there uniquely (Tables 5.6, 5.8). Thirteen species were found in the cochas, of which the unidentified serrasalmine is probably an artifact. Only a single unique species was found in the lake, *Odontostilbe* sp. 1.

The similarity among habitats (Table 5.6) are all higher than would be expected due to random similarity (Figs. 5.2, 5.3). The lowest similarity, 53.8 between cochas and tributaries, is still five standard deviations ($P<0.001$, two tailed) above mean random similarity. As with the analysis of sub-basins we reject homogeneity and conclude that there is a correlation or dependence among the macrohabitats.

A non-hierarchical dendrogram diagram summarizes the maximal pattern of similarity among the macrohabitats (Fig. 5.7). With the exception of lake habitats, the highest similarities are to riverine habitats, which probably results from seasonal cycles of flooding.

Tributaries are the most divergent habitats, but they share 34 species uniquely with rivers. Most of these shared species were found in the lower reaches or the slightly larger habitats within the tributaries, comprising species of different sizes, ecologies, and feeding guilds, such as *Gymnotus* cf. *anguillaris*, *Hypoptopoma, Microglanis, Creagrutus, Piabucus, Tatia, Triportheus* and *Tyttocharax*. Surprisingly, not a single species of cichlid is in that list.

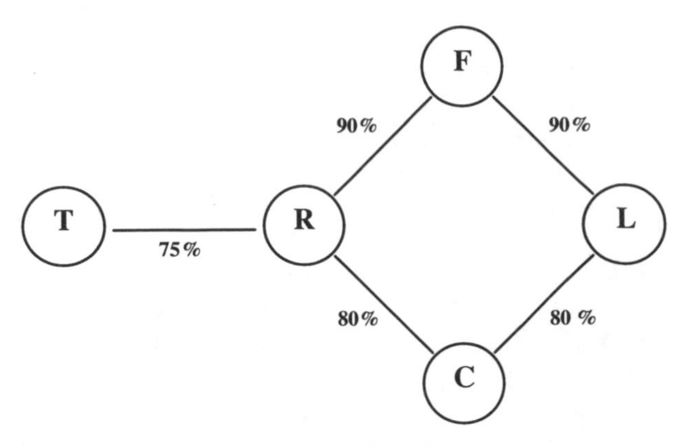

Figure 5.7. The network among macrohabitats abbreviated within the circles based upon an overlapping, non-hierarchical cluster analysis of the percent similarities, Simpson's index, among the macrohabitats. The values shown are the percent similarities between the macrohabitats connected by the network. Abbreviations: C - cocha; F -flooded forest; L - lake; R - river; T - tributary.

Despite the fact that we found just over 100 species in cochas (Tables 5.6, 5.8), 22 were shared uniquely with rivers. These species included a number that prefer quieter water, among rooted vegetation and with many sticks or logs, such as *Abramites*, *Crenicara*, *Eigenmannia*, *Hypostomus*, *Nannostomus*, *Parotocinclus*, and *Mesonauta*.

The extremely high, 90% similarity of flooded forests to rivers and to lakes is not due to joint possession of unique taxa. Rather, almost all of the species that were found in the forest habitats had broad distributions and were almost always found in either rivers or lakes. Of the 40 species found in rivers, 12 are distributed everywhere, 2 are unique and the other 26 are found in a variety of habitats but always in rivers or lakes.

The distribution of species among the macrohabitats was analyzed to determine if there was an overall pattern. The degree of disorder was found to be more orderly than the resulting distribution from 500 Monte Carlo simulations that produced a mean temperature of 50.62 with a standard deviation of 3.16. The probability of obtaining the observed temperature, 21.15, at random is 6.4×10^{-21}. Thus, there is significant structure to the distribution of species among macrohabitats. Unlike the analysis of sub-basins there is no way to really test whether the distribution is clinal or nested because there is no logical adjacency relationship among the habitats other than that due to flooding. However, that the similarities are so high even for lakes and cochas with tributaries may imply a nested pattern. We will treat the distribution as dependent and significantly patterned.

Table 5.8. Fish species found living only in tributaries (n=29), cochas (n=13), and in all (n= 12) macrohabitats with the exception of flooded forests.

Tributary	Cocha
Ancistrus sp. 1	*Aequidens* cf. *tetramerus*
Aphyocharax alburnus	*Charax gibbosus*
Bryconamericus cf. *caucanus*	Cheirodontinae sp.
Bryconamericus cf. *peruanus*	*Corydoras hastatus*
Characidium sp. 2	*Curimatella alburna*
Chrysobrycon sp. 1	*Curimatella meyeri*
Chrysobrycon sp. 2	*Cyphocharax* sp.
Corydoras sp.	*Dysichthys bifidus*
Creagrutus sp. 1	*Electrophorus electricus*
Cynopotamus gouldingi	*Microgeophagus altispinosa*
Gephyrocharax sp.	*Moenkhausia* sp. 6
Gymnotus cf. *coatesi*	*Odontostilbe* sp. 2
Heptapterus longior	*Potamorhina altamazonica*
Heptapterus sp.	Serrasalminae sp.
Hydrolycus pectoralis	**All Macrohabitats**
Hyphessobrycon agulha	*Astyanax* cf. *abramis*
Hyphessobrycon cf. *gracilior*	*Corydoras trilineatus*
Imparfinis sp.	*Crenicichla* cf. *heckelii*
Leporinus friderici	*Ctenobrycon spilurus*
Loricariidae sp.	*Cyphocharax spiluropsis*
Microschemobrycon geisleri	*Hoplias malabaricus*
Pimelodus armatus	*Moenkhausia dichroura*
Potamorrhaphis sp.	*Moenkhausia sanctaefilomenae*
Rhamdia sp.	*Odontostilbe fugitiva*
Rhaphiodon vulpinus	*Odontostilbe hasemani*
Schizodon fasciatum	*Odontostilbe* cf. *paraguayensis*
Sorubim lima	*Steindachnerina dobula*
Tyttocharax sp. nov.	
Tyttocharax tambopatensis	

III. Water

Three types of water were recognized in the Tahuamanu-Manuripi region: black, white and turbid. They contained the following number of species: black – 188; white – 274; and turbid – 246. Blackwaters also contained 36 species not found in other water types that were almost equally distributed among river, cocha and flooded forest habitats (Table 5.9). Whitewaters contributed another 22 species that were not found elsewhere (Table 5.9).

The similarity among water types was exceedingly high, >90% for white and turbid waters, and 76-80% for those with black water. These values are within the distribution of random similarities, and we fail to reject the null hypotheses of homogeneity ($P > 0.05$, two tailed).

Similarly, the matrix of species distributions over water types had a degree of disorder, T = 41.63, that was not significantly different (P = 0.243) from the mean of the Monte Carlo simulations, T = 44.64. Thus, we conclude that there is no significant distributional effect due to water type.

Failure to find a significant effect due to water is somewhat surprising given the information available relating changes in faunal assemblages to water chemistry, particularly to blackwaters (Lowe-McConnell, 1987; Ibarra and Stewart, 1989; Cox Fernandes, 1995). However, the sampling of water types among macrohabitats and among sub-basins was too coarse to permit significant effects to be analyzed. That is, so many species are found in each of the water types that rejecting homogeneity is almost impossible.

DISCUSSION

This report has dealt with the distribution of freshwater fishes in a relatively restricted region of the Upper Río Orthon basin, Pando, Bolivia. The basis for the distributional data came from a survey in September 1996, providing point source information from a rapid assessment. Despite the rapid nature of the survey, we have documented perhaps the richest ichthyological region within all of Bolivia (Chernoff et. al., 1999). Yet from a conservation perspective, we have attempted to develop a basis for understanding the dimensions of the distributions of the fishes in the Tahuamanu-Manuripi region in order to assess the impact of current or proposed threats.

Threats to the environment of this region (e.g., logging) will have vastly different impacts depending upon the distribution patterns of the fishes. For example, if all the species found within a region were homogeneously distributed, then environmental destruction to one part of the region or to one type of habitat might not diminish the number of species. However, departure from homogeneity and randomness, whether increasing or decreasing patterns of similarity, could more easily result in loss of species. We first summarize the results above in relation to the three null hypotheses that the distributions of fishes in the Manuripi-Tahuamanu are homogeneous with respect to: i) sub-basins; ii) macrohabitat; and iii) water type. Then we use that information in a predictive fashion relative to the current environmental threats to the region. We will attempt to provide lists of taxa as profiles for the assemblages that might survive under certain conditions.

Table 5.9. Fish species found in blackwater (n=36) and whitewater (n=22) habitats.

Black Water	White Water
Acanthodoras cataphractus	*Anodus elongatus*
Agamyxis pectinifrons	*Apistogramma* sp. 3
Ancistrus sp. 4	*Astronotus crassipinnis*
Apistogramma sp. 4	*Auchenipterus* cf. *nuchalis*
Astrodoras asterifrons	*Cheirodon* sp. 2
Brachyhypopomus pinnicaudatus	*Cynodon gibbus*
Cheirodontinae sp.	*Duopalatinus* cf. *malarmo*
Cichla cf. *monoculus*	*Heptapterus* sp.
Cichlasoma severum	*Imparfinis* sp.
Corydoras cf. *napoensis*	Loricariidae sp.
Crenicara cf. *unctulata*	*Moenkhausia* sp. 1
Curimatella alburna	*Odontostilbe* sp. 1
Dianema longibarbis	*Plagioscion squamosissimus*
Glyptoperichthys lituratus	*Pristobrycon* sp.
Hemigrammus cf. *pretoensis*	*Pseudodoras niger*
Hemigrammus cf. *unilineatus*	*Pseudohemiodon* sp. 1
Hyphessobrycon cf. *anisitsi*	*Pseudostegophilus nemurus*
Hyphessobrycon cf. *gracilior*	*Roeboides* cf. *myersi*
Hypopygus lepturus	*Serrasalmus* cf. *marginatus*
Laemolyta sp.	*Serrasalmus marginatus*
Metynnis luna	*Serrasalmus* sp.
Microgeophagus altispinosa	Tetragonopterinae sp.
Moenkhausia cf. *comma*	
Moenkhausia sp. 6	
Nannostomus trifasciatus	
Pimelodidae sp.	
Platydoras costatus	
Pseudoplatystoma fasciatum	
Pyrrhulina australe	
Rivulus sp.	
Satanoperca cf. *acuticeps*	
Scoloplax cf. *dicra*	
Serrasalminae sp.	
Tatia aulopygia	
Trachelyopterus cf. *galeatus*	
Tridentopsis pearsoni	

Distributional Patterns and Null Hypotheses

The results of similarity patterns and their tests against null hypotheses of homogeneity allowed us to reject the null hypotheses for both sub-basin and macrohabitat effects but not for water type. For both sub-basins and macrohabitats similarities were greater than to be expected at random ($P<0.01$), indicating interdependency on variables associated with geography or environment. When tested further, we were able to reject ($P<<0.001$) the null hypotheses of random pattern using measures of disorder in the sub-basin or macrohabitat by species matrices. In the case of sub-basins, patterns of similarity and patterns of shared species, both exclusive and ubiquitous (Fig. 5.6) allowed us to discount smooth faunal turnover as the pattern. Rather, the sub-basin effect is closer to that of nested subsets with idiosyncratic distributions or even faunal shifts. For macrohabitats, nested patterns are most consistent with a fauna being derived from (i.e., moving through) riverine habitats on seasonal cycles of flooding.

The results are consistent with published literature on the strong geographic and macrohabitat related effects on freshwater fish distributions (Balon and Stewart, 1983; Balon et al., 1986; Hawkes et al., 1986; Matthews, 1986; Ibarra and Stewart, 1989; Cox Fernandes, 1995). The accumulations of species in regions, even small ones, are discontinuous and act more like sets of islands than as uniform habitat. Diversity of fish species in the Tahuamanu-Manuripi region is a non-random interaction of environmental quality and habitat diversity distributed through these river basins. Only 33 out of 313 species are found in all sub-basins.

The Manuripi had the largest fauna and the most complete sampling of macrohabitats and water types; only tributary macrohabitats were not present in the area sampled. This overall diversity in the Río Manuripi was exceptional with 220 species, but, importantly, 78 were found only there and 24 of these species represent new records for Bolivia. Whereas the Río Nareuda was dominated by tributary and river habitats with only a single blackwater locality, it still had 166 species, but only 25 exclusively within the region of which eight were new records for Bolivia. But the heterogeneity and complexity of the factors affecting distribution is best realized by comparing the Upper and Lower Río Nareuda. Though their species numbers are almost the same (n=116, 133, respectively) their similarity is only 56%. That is, they only share 63 species in common and there is a 53 species turnover between the Lower and Upper Nareuda. Only 10 species were unique to the river. The Upper Nareuda was actually most similar to the Upper Tahuamanu and then to the Manuripi (Table 5.2).

These few examples demonstrate the complexity of the pattern. Because of the nature of rapid surveys and the requirement for replicates of environmental variables, we are forced to lump habitats into broader categories that might otherwise be more finely divided. For example, the Middle Río Tahuamanu contains many rocky cascades (cachoeras), boulders and outcrops, habitats not discovered in other rivers. The Middle Río Tahuamanu shares about half of its fauna with the Río Manuripi but still had 19 species not captured elsewhere within the study area of which 8 are new records for Bolivia (Table 5.4).

The diversity of species in the Tahuamanu-Manuripi region does not show an up- to downstream increase in diversity as predicted by a number of North American studies (e.g., Gorman and Karr, 1978). Neither was this phenomenon observed in the Río Napo of Ecuador in the classic study by Ibarra and Stewart (1989). It is possible, however, that our distances were not sufficient to ascertain this phenomenon (which was surmised by Ibarra and Stewart as well). However, we raise the possibility here that within lowland, Amazonian river systems there may not be as marked a drop in diversity as one proceeds upstream. We will be testing this notion with future sampling studies.

In conclusion, our results agree in general with those of Ibarra and Stewart (1989) for the Río Napo. Our geographic and macrohabitat analyses demonstrate that the fish assemblages do not change smoothly as one moves along environmental gradients. Rather there are some zones of minimal change (e.g., from the Manuripi to the Middle Río Tahuamanu or from rivers into flooded forests) and then boundaries or regions of more abrupt but non-random faunal turnover (e.g., ascending the Río Nareuda from the Middle Tahuamanu). Though our region of study is more restricted than those considered by other authors, our results on non-random patterns of zonation accord well with their findings (Campos, 1985; Balon et al., 1986; Ibarra and Stewart, 1989; Cox Fernandes, 1995). We will now use these non-random patterns in an analysis of environmental threats.

Analysis of Threats

The major threats facing the region of the Tahuamanu-Manuripi region is logging and human development of the area: road construction, land clearing, cattle ranching. Because the distribution of fishes is not homogeneous but in a zoned pattern, these destructive activities will impact the habitats within the region in slightly different ways.

In our analysis we believe that those species that are most widely distributed across habitats, sub-basins, water types and even in other drainages may be most resistant to environmental perturbations. The 33 species (Table 5.3) that were in all sub-basins and live in more than one habitat may be the most resilient and, we predict, would survive environmental degradation to some parts of the river basin. Of these taxa, eight species represent new records for Bolivia.

The majority of the diversity of freshwater fishes are found in the Río Manuripi sub-basin (220 species) or in rivers (223 species), each of which contain remarkably similar numbers of unique species for the region (78 and 91 respectively) and new records for Bolivia (24 and 30 respectively). The overlap of the two is high, containing 170 species which equals 54% of the fishes found in the Tahuamanu-Manuripi region. Although 101 species were found in the cochas, lakes and flooded areas of the Manuripi River basin, only 17 of these were not also found in the main river. To try to use these data to conclude that the cochas, lakes and flooded areas do not contribute much to the biodiversity within the Manuripi system would be incorrect. Many of the species found both in the rivers and the quieter water bodies depend upon these quieter habitats (Goulding, 1980; Lowe-McConnell, 1987; Machado-Allison, 1987) for much of their lives (e.g., the cichlids, *Apistogramma* spp., *Corydoras* spp., *Parotocinclus* sp., etc.) and their food (e.g., *Hydrolycus pectoralis, Potamorhina altamazonica,* and *Curimatella* sp.).

Environmental impacts to the Manuripi region, via land transformation, road construction and logging could impact the main river, the flooded areas (varzea) and the cochas and lakes independently. Human activities were beginning to place cochas, lakes and varzea at risk above Puerto Rico. Forests and varzeas had been cleared, altering the margins of the habitats by activities such as replacement of natural vegetation and trees with what appeared to be a form of "lawn grass". Such intrusions upon the natural landscape appeared to be resulting in increased sedimentation, unstable banks and also human pollution through trash. Continued perturbations would likely place a number of the 101 species dependent upon these flooded zones at risk. We estimate that more than 50% of these species might disappear from the Manuripi, a percentage that would include almost all the species of cichlids, many ornamentals, etc. Also included would be many commercially valuable species that use the areas for reproduction, such as pacus, *Mylossoma duriventris*, payaras, *Hydrolycus pectoralis*, piranhas, *Serrasalmus* sp. and bocachicos, *Prochilodus nigricans*. Those species that we estimate would survive extensive destruction would be those that depend more heavily upon the rivers for food, reproduction and survival, only making occasional forays into the flooded zones (e.g., some small catfishes, *Pimelodus* spp., *Pimelodella* sp., tetras, *Moenkhausia dichroura, Aphyocharax alburnus*).

Because the fishes are not homogeneously distributed and display strong zonal or nested patterns relative to sub-basin geography and macrohabitat, we have identified from the patterns of similarity and from the analysis of matrix disorder the likely sources for the maintenance of or potential recolonization of species that would disappear under the scenario of extensive destruction of cochas, lakes and flooded areas in the Manuripi. The remaining fauna would primarily comprise the 33 widespread species (Table 5.3, plus 21 species shared uniquely with the Middle Río Tahuamanu and the Lower Río Nareuda (Table 5.10) and 11 species shared uniquely with the Upper Río Tahuamanu or the Upper Río Nareuda (Table 5.10). Thus, of the 101 species found prior to the scenario of habitat destruction, only 67 would remain within the entire area. That would represent a net loss of 34 species of which 12 represent new records for Bolivia. The other sub-basins are potential sources for recolonization of the Manuripi flooded areas, but whether fishes move actively across both the geographic and macrohabitat zones would have to be studied.

The most threatened habitat of all are the tributary habitats, located in the upper areas of the Río Tahuamanu and Río Nareuda. In the tributaries we found 128 species of which 29 were found there uniquely, the latter quantity containing 11 new records for Bolivia. We found the logging, road construction and land conversion for cattle ranching to be extensive in this region. Many of the small garapes lacked shade or native vegetation along their banks. Large deposits of mud from siltation were on the increase in these habitats. Runoff from recently burned or sprayed fields is toxic to freshwater fishes. We found a number of small streams in forested habitat, but downstream from recently burned and cleared fields, lacking any fishes.

We predict that continued destruction to these highly vulnerable habitats will result in the following changes. With sufficient perturbation, 24 of the unique species (Table 5.8, not including *Sorubim, Hydrolycus, Leporinus, Rhaphiodon*, and *Potamorrhaphis*) could likely disappear from the Tahuamanu-Manuripi region. The base fauna would comprise the 33 species found in all sub-basins and the 12 species found in all macrohabitats (Tables 5.3, 5.8); these are the likely survivors or recolonists. Of the 54 species that inhabit tributaries but are not ubiquitously distributed (Table 5.11), 64% are found in the Río Manuripi, either alone (n=8) or in combination with Middle Río Tahuamanu or the Lower Río Nareuda (n=27). This derives from the non-random pattern in which the upper tributaries shared a relatively large portion of their fauna with the Río Manuripi. This means that if habitats in the Río Manuripi were conserved and maintained then, under the current scenario of degradation to the upper tributaries, at least 80 of the 128 species found in these habitats would still exist within the Tahuamanu-Manuripi region. Furthermore, if the degradation does not cascade downstream to the Lower Río Nareuda or the Middle Río Tahuamanu, an additional 27 species of the upper tributaries could be maintained that are not found within the Río Manuripi.

Maintenance of the Middle Río Tahuamanu and Lower Río Nareuda is critical not only for the fauna that they share with their tributaries but also for the preservation of many unique species (Tables 5.4, 5.5). Unfortunately, this is not a likely scenario. Deforestation, including modification of the river banks, was extensive above Filadelfia on the Río Tahuamanu. Habitat destruction was projected to continue, actions that will leave the upper tributaries of the Middle to Upper Río Nareuda and the Upper Tahuamanu as refuges for the 54 species that they share downstream (Fig. 5.6). These upper regions of the Nareuda and Tahuamanu could, thus become islands. Unfortunately many of the taxa listed in Table 5.11, and certainly the larger fishes (e.g., *Rhaphiodon*, *Prochilodus*, the piranhas) will probably not survive in limited habitat-islands.

Table 5.10. Potential recolonists for quiet water habitats in Manuripi. These fish species were captured at more than one locality. The most likely source of recolonization was from populations living in the Middle Río Tahuamanu or the Lower Río Nareuda (n=21). If longer distance recolonization is possible, then additional colonists may have come from the Upper Río Tahuamanu and the Upper Río Nareuda (n=11).

M. Tahuamanu or L. Nareuda	U. Tahuamanu or U. Nareuda
Abramites hypselonotus	*Aequidens* sp. 1
Astyanax cf. *abramis*	*Corydoras acutus*
Carnegiella myersi	*Corydoras trilineatus*
Cheirodon fugitiva	*Dysichthys* cf. *amazonicus*
Cochliodon cf. *cochliodon*	*Gasteropelecus sternicla*
Corydoras cf. *loretoensis*	*Hypoptopoma joberti*
Creagrutus sp. 1	*Megalechis thoracatus*
Ctenobrycon spilurus	*Phenacogaster* cf. *pectinatus*
Cyphocharax spiluropsis	*Phenacogaster* sp. 2
Hoplias malabaricus	*Rineloricaria lanceolata*
Moenkhausia colletti	*Rineloricaria* sp.
Moenkhausia dichroura	
Moenkhausia sanctaefilomenae	
Ochmacanthus cf. *alternus*	
Odontostilbe paraguayensis	
Otocinclus mariae	
Pimelodella gracilis	
Pimelodella cf. *itapucuruensis*	
Prionobrama filigera	
Steindachnerina dobula	
Sturisoma nigrirostrum	

CONCLUSIONS AND RECOMMENDATIONS

The role of rapid assessment programs is to provide a critical but brief snapshot of the composition and structure of ecosystems. Because of the imminent threats to ecosystems and their organisms across the globe (Myers, 1988, 1990; Sisk et al., 1994), a strategy of rapid assessments across landscapes can target the best and most valuable habitats for immediate conservation and detailed study. To provide maximal value to the process, the rapid assessments should provide not only surveys of the organisms and vital environmental correlates but also hypotheses about the community structure within the region as well as resilience to environmental threats.

This task is particularly daunting in application to aquatic ecosystems within tropical South America because of the often unappreciated dynamics and complexity within these ecosystems (e.g., Goulding 1980, 1981; Goulding et al., 1988; Machado-Allison, 1990; Barthem and Goulding, 1997). The wonderful perspectives presented on the importance of and challenges facing conservation of aquatic ecosystems (Naiman et al., 1995; Pringle, 1997), are often based exclusively upon relatively simple temperate systems. Human development strategies (e.g., Almeida and Campari, 1995) within the basins of the principal rivers of South America often ignore completely the impacts to aquatic ecosystems.

The purpose of this paper has been to evaluate the patterns of similarities for the distributions of species among sub-basins, macrohabitats and water types in the Upper Río Orthon basin of Bolivia – here referred to as the Tahuamanu-Manuripi region. This area was discovered to be an unrecognized hotspot of ichthyological diversity (Chernoff et al., 1999) and, therefore, is a high priority for immediate conservation. A preliminary examination of the structure of the fishes found within the region is critical as a primary basis for designing a conservation strategy in relation to current and impending environmental threats.

The methods described and employed are designed to test the null hypotheses of homogeneity of distribution of the fishes (or any organisms) in relation to geographic or environmental variables. Non-homogeneous, patterned distributions require different conservation strategies than does the situation where organisms are homogeneously distributed (e.g., Atmar and Patterson, 1993; Christensen, 1997). Our methods are appropriately conservative and robust, and have the advantage in that we can determine whether the similarity is greater than or less than those expected at random.

This is the first time that we can document such methods applied to point-source data – resulting from a single survey – in order to estimate patterning within an assemblage of species. This in itself is not a conservative

Table 5.11. Tributary fish fauna not including unique species or those found among all sub-basins or macrohabitats. MT = Middle Tahuamanu; LN = Lower Nareuda; MAN = Manuripi; R = River; L = Lake; C = Cocha.

Species	Sub-basin and Macrohabitat	Species	Sub-basin and Macrohabitat
Aequidens cf. *paraguayensis*	MT, LN, R	*Hemiodontichthys acipenserinus*	LN, MAN, R, C
Aequidens sp. 1	LN, R	*Hemisorubim platyrhynchos*	MAN, R
Ageneiosus sp.	MT, R	*Homodiaetus* sp.	MAN, R
Ancistrus sp. 3	LN, R	*Hoplosternum thoracatus*	MAN, C
Aphanotorulus frankei	MT, LN, R	*Hypoptopoma* sp.	LN, MAN, R
Aphyocharax dentatus	MT, MAN, R	*Hypostomus* sp. 1	LN, MAN, R
Apistogramma linkei	LN, MAN, R	*Knodus* sp.	Everywhere, R
Apistogramma sp. 1	MAN, R, L	*Megalonema* sp.	MT, MAN, R
Apistogramma sp. 2	MT, LN, MAN, R	*Microglanis* sp.	LN, R
Apteronotus albifrons	LN, MAN, R	*Moenkhausia* cf. *lepidura*	Everywhere, R, C
Brachychalcinus copei	LN, R	*Mylossoma duriventris*	MAN, R
Bryconamericus sp.	MT, R	*Odontostilbe paraguayensis*	MT, LN, MAN, R, C
Bunocephalus cf. *aleuropsis*	MT, LN, R	*Phenacogaster* sp. 1	MT, MAN, R, C
Bunocephalus cf. *amazonicus*	MT, LN, MAN, R, C	*Phenacogaster* sp. 2	LN, MAN, R, C
Bunocephalus cf. *depressus*	LN, R	*Phenacogaster* ? sp.	MT, R
Characidium sp. 1	MT, LN, R	*Piabucus melanostomus*	MAN, R
Corydoras aeneus	MT, R	*Pimelodella cristata*	LN, MAN, R, C
Creagrutus sp. 3	MT, R	*Prochilodus* cf. *nigricans*	LN, MAN, R, C
Crenicichla sp. 2	LN, MAN, R, C	*Pyrrhulina vittata*	MT, MAN, R, C
Doras cf. *carinatus*	MT, MAN, R	*Serrasalmus rhombeus*	MAN, R
Eigenmannia macrops	MT, MAN, R	*Steindachnerina guentheri*	MT, LN, R, C
Farlowella cf. *oxyrryncha*	MT, MAN, R, C	*Steindachnerina* sp.	MT, LN, R, C
Farlowella sp. 1	MT, LN, MAN, R	*Sternopygus macrurus*	MAN, R
Galeocharax gulo	MT, R	*Synbranchus marmoratus*	MAN, R
Hemigrammus ocellifer	MT, LN, MAN, R, C	*Tyttocharax madeirae*	MT, LN, R

approach to community ecology. But again, we believe that rapid assessment programs must provide beginning insights into conservation plans to protect ecosystems and biodiversity to the maximum extent possible. We are persuaded by the elegant argumentation of Shrade-Frechette and McCoy (1993). The ethics of conservation require that we risk a type-I error (rejection of a true null hypothesis) rather than a type-II error (failure to reject a false null hypothesis). In the case presented here this means that it is better to err on the side of recognizing structure within a biological assemblage. The conservation plan is more stringent than that required to conserve a homogeneously distributed community (Christensen, 1997; Barrett and Barrett, 1997). We, therefore, conclude and recommend the following:

1. The freshwater fishes are non-randomly distributed with respect to sub-basins and macrohabitats within the Tahuamanu-Manuripi region. The pattern mostly closely approximates a system of zonation with distinct boundaries. At the boundaries there can be 50% turnover of species.

2. The freshwater fishes are homogeneously distributed with respect to water types. Even so, there are 39 species that were found only in black (acid) waters.

3. It is critical to designate two core areas within the Tahuamanu-Manuripi region as core conservation areas: the Manuripi and the Nareuda. Together they represent more than 75% of the diversity within the region.

4. Both of these areas are at risk because of the narrow flood zone within the Upper Río Orthon basin. The narrow nature of the flood zone brings human activities closer to, and is now infringing upon, the varzeas, cochas and flooded areas. More than 80% of the fishes depend upon the flooded areas for reproduction, nursery grounds or critical food getting.

5. The streams crossing terra firme zones, such as the garapes, are the most endangered habitats. Their degradation could imperil 30% of the species and 11 species that are new to Bolivia, including a number of highly valuable ornamental fishes (e.g., *Chrysobrycon*).

6. Any degradation to the Manuripi system will require setting a number of alternate zones, such as the Middle Río Tahuamanu and the Lower Nareuda as new core zones.

LITERATURE CITED

Almeida, A. L. O. de, and J. S. Campari. 1995. Sustainable development of the Brazilian Amazon. Oxford University Press, New York. 189 pp.

Atmar, W., and B. D. Patterson. 1993. The measure of order and disorder in the distribution of species found in fragmented habitat. Oecologia 96: 373-382.

Balon, E. K., S. S. Crawford, and A. Lelek. 1986. Fish communities of the upper Danube River (Germany, Austria) prior to the new Rhein-Main-Donau connection. Environmental Biology of Fishes 15: 243-271.

Balon, E. K., and D. J. Stewart. 1983. Fish assemblages in a river with unusual gradient (Luongo, Africa – Zaire system), reflections on river zonation and description of another new species. Environmental Biology of Fishes 9: 225-252.

Barrett, N. E., and J. P. Barrett. 1997. Reserve design and the new conservation theory. *In* Pickett, S.T.A., R. S. Ostfield, M. Schachak, and G. E. Likens (eds.). The ecological basis of conservation: heterogeneity, ecosystems, and biodiversity. Pp. 236-251. Chapman and Hall, New York.

Barthem, R., and M. Goulding. 1997. The catfish connection: ecology, migration, and conservation of Amazon predators. Columbia University Press, New York. 144 pp.

Campos, H. 1985. Distribution of the fishes in the Andean rivers in the south of Chile. Archiv. Hydrobiol. 104: 169-191.

Chernoff, B., P.W. Willink, J. Sarmiento, S. Barrera, A. Machado-Allison, N. Menezes, and H. Ortega. 1999. Fishes of the Ríos Tahuamanu, Manuripi and Nareuda, Depto. Pando, Bolivia: diversity, distribution, critical habitats and economic value. *In* Chernoff, B., and P.W. Willink (eds.). A biological assessment of the aquatic ecosystems of the Upper Río Orthon basin, Pando, Bolivia. Pp. 39-46. Bulletin of Biological Assessment No. 15. Conservation International, Washington, D.C.

Christensen, N. L., Jr. 1997. Managing for heterogeneity and complexity on dynamic landscapes. *In* Pickett, S.T.A., R. S. Ostfield, M. Schachak, and G. E. Likens (eds.). The ecological basis of conservation: heterogeneity, ecosystems, and biodiversity. Pp. 167-186. Chapman and Hall, New York.

Cox Fernandes, C. 1995. Diversity, distribution and community structure of electric fishes (Gymnotiformes) in the channels of the Amazon River system, Brasil. Unpublished Ph.D. Dissertation, Duke University. 394 pp.

Goulding, M. 1980. The fishes and the forests: explorations in Amazonian natural history. University of California Press, Los Angeles. 280 pp.

Goulding, M. 1981. Man and fisheries on an Amazon frontier. Dr. W Junk Pub., Boston. 137 pp.

Goulding, M., M. L. Carvalho, and E. G. Ferreira. 1988. Río Negro: rich life in poor water: Amazonian diversity and foodchain ecology as seen through fish communities. SPB Academic Publishing, The Hague, Netherlands. 200 pp.

Gorman, O. T., and J. R. Karr. 1978. Habitat structure and stream fish communities. Ecology 59: 507-515.

Hawkes, C. L., D. L. Miller, and W. G. Layher. 1986. Fish ecoregions of Kansas: stream fish assemblage patterns and associated environmental correlates. Environmental Biology of Fishes 17: 267-279.

Ibarra, M., and D. J. Stewart. 1989. Longitudinal zonation of sandy beach fishes in the Napo River basin, eastern Ecuador. Copeia 1989: 364-381.

Lowe-McConnell, R. H. 1987. Ecological studies in tropical fish communities. Cambridge University Press, New York. 382 pp.

Machado-Allison, A. 1987. Los peces de los llanos de Venezuela: un ensayo sobre su historia natural. Universidad Central de Venezuela Consejo de Desarollo Cientifico y Humanistico, Caracas. 141 pp.

Machado-Allison, A., J. Sarmiento, P.W. Willink, N. Menezes, H. Ortega, and S. Barrera. 1999. Diversity and abundance of fishes and habitats in the Río Tahuamanu and Río Manuripi basins. *In* Chernoff, B., and P.W. Willink (eds.). A biological assessment of the aquatic ecosystems of the Upper Río Orthon basin, Pando, Bolivia. Pp. 47-50. Bulletin of Biological Assessment No. 15. Conservation International, Washington, D.C.

Matthews, W. J. 1986. Fish faunal "breaks" and stream order in the eastern and central United States. Environmental Biology of Fishes 17: 81-92.

Myers, N. 1988. Threatened biotas: "hotspots" in tropical forests. The Environmentalist 8: 1-20.

Myers, N. 1990. The biodiversity challenge: expanded hot-spots analysis. The Environmentalist 10: 243-256.

Naiman, R. J., J. J. Magnusin, D. M. McKnight, and J. A. Stanford. 1995. The Freshwater Imperative. Island Press, Washington, D.C. 165 pp.

Pringle, C. M. 1997. Expanding scientific research programs to address conservation challenges in freshwater ecosystems. *In* Pickett, S.T.A., R. S. Ostfield, M. Schachak, and G. E. Likens (eds.). The ecological basis of conservation: heterogeneity, ecosystems, and biodiversity. Pp. 305-319. Chapman and Hall, New York. 466 pp.

Shrader-Frechette and E. D. McCoy. 1993. Method in Ecology. Cambridge University Press, New York. 328 pp.

Sisk, T. D., A. E. Launer, K. R. Switky, and P. R. Ehrlich. 1994. Identifying extinction threats. BioScience 44: 592-604.

Sokal, R. R., and N. L. Oden. 1976. Spatial correlation in biology: I. methodology. Biological Journal of the Linnean Society 10:199-228.

CHAPTER 6

LANDSCAPE-BASED GENETIC ASSESSMENT OF THE FISHES OF THE RIOS TAHUAMANU AND MANURIPI BASINS, BOLIVIA

Theresa M. Bert

ABSTRACT

As a unique measure of biodiversity independent of species diversity and because the existing taxonomy may not reflect the underlying genetic diversity of the fish species in the Río Manuripi drainage, we included an assessment of the genetic diversity, and genetic population structure of selected freshwater fish during the AquaRAP expedition to the Pando, Bolivia in September 1996. Tissue or whole-body samples of 500 fish were collected from approximately 50 genera of fishes from the Río Manuripi drainage for subsequent analysis of population genetic structure or phylogenetic relationships. Samples were frozen in liquid nitrogen or preserved in 100% ethanol. The bodies of fish from which we took tissue samples were preserved as voucher specimens in formalin. For most putative species, one to a few individuals were sampled; for a few species, multiple collections suitable for population genetic analysis were taken. We are in the process of genetically analyzing these samples. The results of that work will be available as a supplement to this document. The structure of the Río Manuripi drainage system is highly heterogeneous and should provide ample opportunity for selection to operate and speciation processes to occur. The diversity among streams and cochas in the relative abundance of common species suggests that selection is strong in its action on at least some fish species and that it may contribute to maintenance of a relatively high level of genetic diversity within species. From a genetic perspective, perhaps the most important river basin to preserve is the Manuripi. Our preliminary evaluation suggests that it is important to preserve both the water quality and the physiography of this basin. This, of course, necessitates the preservation, at least to some degree, of the entire drainage system.

INTRODUCTION

Multiple, independent measures of a region's biotic systems provide greater insight into its history and processes and a better basis for predicting future biodiversity (McKinney et al., 1998) than do unilateral measures. Genetic analysis can provide a unique, independent assessment of the conservation value of a region by facilitating novel insights relevant to conservation (Avise, 1996). A faunal or floral component may be genetically valuable regardless of the number of species or proportion of endemism in the component. For example, a region may include an area of numerous hybrid zones, where gene complexes are recombining in novel ways, or a region may harbor an unusually high percentage of ancient lineages and thereby serve as a genetic repository for ancestral genotypes. Alternatively, a region may serve as a potential source of evolutionary importance because many species maintain unusually high genetic variation. Both hybrid zones and regions possessing representatives of species-rich clades are thought to be sites of ongoing evolution and speciation (Harrison, 1990; Erwin, 1991) and are important for the generation of future biodiversity (Avise, 1996). Moreover, the co-occurrence of multiple hybrid zones in an area can be a strong indicator of the presence of an ecotone because hybrid zones tend to both be generated and to collect in ecotones (Barton, 1985; Moore and Price, 1993). Such areas are sources of evolutionary novelty and are important to preserve (Smith et al., 1997). Regions possessing exceptionally high proportions of species with ancient genomes make a disproportionate contribution to the world's pool of evolutionary diversity (Vane-Wright et al., 1991).

Human activities have impacted freshwater ecosystems to a greater extent than terrestrial ecosystems (Vitousek et al. 1997). As high as they are, losses of species far underestimate the magnitude of the loss of genetic variation (Vitousek et al., 1997). In addition, the existing taxonomy for a group may not reflect the underlying genetic diversity (Moritz, 1994). Because genetic variation is the raw material of natural selection (Fisher, 1930), biodiversity ultimately *is* genetic diversity (Avise, 1996). The long-term production and usefulness of populations depends on preserving genetic variation among and within populations (Currens and Busack 1995). Recognizing this, the Convention on Biodiversity specifically recognized the genetic component of biodiversity and recommended inclusion of genetic diversity considerations in assessments of the conservation value of a region.

We included an evaluation of genetic biodiversity in our complement of protocols for the Río Manuripi drainage AquaRAP assessment. The structure of the Río Manuripi drainage, which includes numerous caños, streams, cochas, and rivers ranging from small and clear (blackwater) to large and sediment-laden (whitewater), and the patterns in the distribution and relative abundance of certain fish species suggest that genetic distinctness and variation are promoted through local selection and stochastic events. Our sampling strategy was designed to allow an assessment of both intraspecific genetic variation among populations and the relative age of lineages and structure of their interrelationships. Here, I present a synopsis of the collections taken on that expedition and a preliminary evaluation of the potential importance of the Río Manuripi drainage for conservation.

METHODS

On this expedition, samples specifically for genetic analysis were collected only for fish. However, the samples of terrestrial plants also collected during this expedition were preserved in alcohol and would be suitable for any DNA-based analysis.

For the fish samples destined for genetic analysis, the overall sampling strategy had two components. To generate a library of tissues for archival as museum collections for species identification and other research, tissue samples were taken from one to a few individuals from many different morphological species. To provide samples for investigating the genetic population structuring of fish in the Nareuda-Tahuamanu-Manuripi river system, a larger number of individuals (10-50) from particular species were sampled. To analyze the population genetic structure of a species, large numbers of individuals must be captured from different locations, preferably from different streams and rivers within a drainage and among drainages. Since we were not sure whether we would be able to capture large numbers of a particular species in more than a single location, we collected large numbers of individuals for several species that could be tentatively identified in the field. For this sampling strategy, the hope was that a species found to be sufficiently abundant for the evaluation of its population genetic structure at a particular site in a particular stream or river would also be found in abundance at other sites in other water masses. Genetic analysis conducted on large samples of a species collected from disparate locations provides information on gene flow and dispersal among streams and rivers.

Two types of preservation were used: freezing in liquid nitrogen and storing in 100% ethanol. Most taxa for which we obtained large samples were preserved only in alcohol. For most taxa for which we sampled only a few individuals, we cut out a section of somatic muscle for freezing and then preserved the rest of the fish in ethanol. For taxa of small fish (usually those < 4 cm total length), we froze some whole individuals and preserved others in ethanol.

More samples were preserved in ethanol than were frozen because our ability to transport frozen tissue samples collected in the field back to their designated long-term storage location (Florida Marine Research Institute) was limited. Regardless of the method used to preserve a sample for genetic analysis, the entire fish was kept as a voucher specimen; either the entire fish was frozen, or a filet was taken for freezing and the remainder of the fish was preserved in ethanol or formalin.

All fish were tentatively identified in the field, most to the generic level. Taxonomic verification of the identifications made in the field was conducted using the voucher specimens. In most cases, 1 species was collected per genus. Here, we report the taxonomic diversity of our collections at the level of the genus.

Logistical complications precluded genetic analysis of the samples prior to completion of this report. Thus in the following sections, the collections made for genetic analysis are first described. Then, based on an understanding of the effects of the ecology and distribution of organisms and the physical dynamics of river systems on population genetic structuring, a general assessment of the genetic conservation value of the Nareuda-Tahuamanu-Manuripi river system is postulated, with emphasis on the fish fauna. The genetic analysis is ongoing; a complete report will be available as a supplement to this report.

Collections

From the Río Nareuda, a total of 43 individuals belonging to 14 genera were sampled. From these fish, 17 frozen samples and 37 ethanol-preserved samples were taken. Most genera were represented putatively by a single species. One to three specimens were sampled for all but two genera. Seven individuals of a *Moenkhausia* species were frozen, and six individuals of a *Bryconamericus* species were preserved in ethanol. The fish were collected from caños entering the river and small sand/mud beaches along the river using beach seines and from the open river using gill nets.

From the Río Tahuamanu, a total of 72 individuals belonging to 26 genera were sampled. From these fish, 53 frozen samples and 70 ethanol-preserved samples were taken.

Thus, all of the fish except two large specimens were preserved in ethanol. Possibly each genus was represented by a single species. One to four specimens were sampled for all genera except two (a Characoid and *Pimelodus*) thought to contain more than one (cryptic) species; for those genera, 10-20 individuals were sampled. Most samples were collected in the main river using beach seines and trawls; two sample collections were made in the Garape Preto, a tributary of the Tahuamanu.

From the Río Manuripi, a total of 385 individuals from 27 genera were sampled. From these fish, 82 frozen samples and 379 ethanol-preserved samples were taken. Thus, the vast majority of the fish were preserved in ethanol, including all of those for which we took frozen samples. Only large specimens (e.g., from the genera *Pteroygoplichthys* and *Serrosalmus*) were preserved in formalin after taking a tissue sample for freezing and/or ethanol preservation. Multiple species were collected for some genera (e.g., *Moenkhausia*, *Pteroygoplichthys*, *Hemigrammus*, *Serrosalmus*); and for a number of morphological species, collections were made from several sites. Sample sizes ranged 1-5 individuals for 67% of the collections, 5-10 individuals for 22% of the collections and 40-100 individuals for the remaining 11% of the collections. All of the collections for which the sample sizes were very large were preserved in ethanol only; no frozen tissue samples were made. Samples were collected along the main river from sand or sand/mud beaches and from open water using beach seines and trawls. Several collections were also made in grassy or muddy lagoons with restricted flow to the main river using gill nets and beach seines and from isolated cochas using beach seines.

In total, samples suitable for use in molecular genetics studies were taken from 500 fish representing approximately 50 genera and over 80 species. Ethanol-preserved samples of 486 of these fish and frozen tissue samples 156 of these fish were taken. Most genera (or species) are represented by only a few individuals per sample. However, samples of *Moenkhausia lepiedura* suitable for analysis at the level of population genetics were collected from multiple streams and rivers, and large samples of several species (e.g., *Hemigrammus* spp., *Pimelodella* spp.) were collected from the Manuripi and its associated aquatic habitats. These samples are suitable not only for gene flow and population structuring estimates within the Nareuda-Tahuamanu-Manuripi River basin but also for population genetic level comparisons of these samples with those that might be collected from neighboring river systems.

RESULTS AND DISCUSSION

Proposed Genetic Conservation Value of the Nareuda-Tahuamanu-Manuripi River System

It is widely recognized that the maintenance of genetic variation is important for the long-term continuation of a species. Genetic variation is necessary to enable species to respond to changing environments. Principally through research related to aquaculture, the negative effects of a loss of genetic variation on a multitude of fitness-related traits have been well documented. In fish, evolution can proceed rapidly in natural settings if ecological conditions are changed and sustained (Reznick et al. 1997). A high level of genetic variability can increase the rate of response to the change and facilitate adaptation to the new environment. Conversely, a loss of genetic variability can lead to a reduction in the ability of a species to adapt to changing environments (Lande, 1988).

Collectively, the Nareuda, Tahuamanu, and Manuripi river basins form a highly heterogeneous habitat that obviously harbors a notably high level of species diversity and intraspecific population substructuring conducive to maintaining genetic variation in species. The apparently narrow niches occupied by numerous species in this river system probably contribute to a high level of population substructuring of those species both within and between river basins. As reflected in the field collections conducted to assess fish species diversity and for genetic analyses, a number of species common at a particular site may be rare or absent from a nearby site but common at other more distant sites. Gene flow between those populations may be restricted, particularly if suitable habitat is uncommon or absent from intervening locations.

Selection may also be a particularly important factor that contributes to the maintenance of genetic variation in fish inhabiting this region. During the "dry season," the receding rivers, particularly the Manuripi, create lakes and ponds (cochas) in old streambeds. These cochas can be isolated from flowing water for months, or perhaps years. In effect, the hydrodynamics of the river system create in the cochas temporary populations that are analogous to peripheral populations and therefore probably subject to the evolutionary processes typically associated with such subpopulations (see Lesica and Allendorf, 1995). These cocha communities tend to be dominated by one or a few species. The species that predominate in the cochas can differ remarkably among cochas, even those in close proximity and superficially similar in habitat composition. The species pool from which these isolated communities were drawn is presumably the common species pool of the river during flood season. Thus, it is reasonable to assume that, aside from stochastic differences, the initial species compositions and relative abundances at the times of isolation from the main river were similar among cochas. Differential strong selection both among and within species must be operating to result in the pronounced differences in species composition and abundance of individuals belonging to a particular species among cochas that we saw after differing but unknown times of isolation of those cochas from the river.

During these periods, selection within species for specific alleles or multiallelic complexes at genetic loci should vary among cochas, depending on such factors as interactions with other species, changes in chemical composition of the water, and benthic habitat. It is possible, even likely, that alleles that are rare in the species as a whole would be under the influence of strong positive selection in these isolated situations; conversely, alleles common in the species could become rare in certain cochas as a result of the same processes. Superimposed on selection would be deviation in the frequencies of alleles of any isolated species population from the allele frequencies of the larger local population inhabiting the river–i.e., sampling error associated with isolating a relatively small percentage of individuals from the local population.

The combined effects of strong selection and sampling error, each of which would be specific to any particular cocha for any species, undoubtedly contribute to the maintenance of genetic variation in those species. Differential selection occurring in the cochas probably results in the development of subpopulations with locally coadapted gene complexes that differ within species among cochas. Thus, this among-population divergence would be an important component of species-level genetic diversity. Moreover, it may contribute to uniqueness of regional populations and to evolutionary adaptability.

During the flood season, these subpopulations would be reunited and recombined and would contribute to the overall gene pool of the species. Population-level gene frequencies could change among years, depending on the relative numbers of individuals from each subpopulation contributing to the local population in a given year. Through this mechanism, alleles may be maintained in the species that would otherwise be eliminated due to directional selection for maladapted alleles or genetic drift in a more homogeneous environment.

In the Nareuda-Tahuamanu-Manuripi system, the most critical habitat to preserve, from a genetic perspective, may be the Manuripi basin and its associated aquatic habitats (streams, cochas, caños, etc.). It is probably particularly important to maintain the physical structure of the Manuripi basin; protection of the Manuripi is needed at the ecosystem level. The creation of dams, channeling of the river, or other changes that result in changes in the physiography of the river basin may have pronounced negative effects on the genetic diversity of the fish species inhabiting the region, and on species composition in general. Changes in biota result from habitat change and land use, and ultimately reduce genetic and species diversity (Chapin et al., 1997). The preservation of whole ecosystems is often the most effective way to sustain genetic diversity because conservation efforts focusing on individual species are expensive and not guaranteed to succeed (Vitousek et al., 1997). Moreover, without detailed knowledge of species interactions in the community, it is difficult to prescribe the relative value of specific species. Species that have small direct effects on ecosystem processes may have large indirect effects if they influence the abundance of species with large direct effects (Chapin et al., 1997).

Throughout the world, the current rate of loss of genetic variability and of species is far above background rates. It is irreversible and represents a global change. However, as high as they are, the losses of species understate the magnitude of the loss of genetic variation (Vitousek et al. 1997). The loss of locally adapted populations within species and of their genetic material reduces the resilience of species to environmental change. Environmental change is inevitable in South America. The maintenance of genetic variability is essential to facilitate species' adaptations to that change. Over time, ecosystems experience an increasingly wide range of conditions (Chapin et al., 1997), and this is exacerbated by anthropogenically related land development. The increasing conversion of undisturbed habitats to those modified for some type of human-associated usage enhances the importance of maintaining genetic diversity to increase the probability of long-term sustainability of ecosystem structure. Preserving genetic diversity could enhance the ability of species to adapt to new environmental conditions and thus influence their survival over ecological or evolutionary time frames (Levin, 1995).

LITERATURE CITED

Avise, J.C. 1996. Introduction: the scope of conservation genetics. *In* Avise, J.C., and J.L. Hamrick (eds.). Conservation Genetics: Case Histories from Nature. Pp. 1-9. Chapman & Hall, New York.

Barton, N. 1985. Analysis of hybrid zones. Annual Review of Ecology and Systematics 16: 113-148.

Chapin, F.S. III, B.H. Walker, R.J. Hobbs, D.U. Hooper, J.H. Lawton, O.E. Sala, and D. Tilman. 1997. Biotic control over the functioning of ecosystems. Science 277: 500-504.

Currens, K.P., and C.A. Busack. 1995. A framework for assessing genetic vulnerability. Fisheries 20: 24-31.

Erwin, T.L. 1991. An evolutionary basis for conservation strategies. Science 253: 750-752.

Fisher, R.A. 1930. The Genetical Theory of Natural Selection. Clarendon Press, Oxford, England.

Harrison, R.G. 1990. Hybrid zones: windows on evolutionary processes. *In* Antonovics, J., and D. Futuyma (eds.). Oxford Surveys in Evolutionary Biology: Volume 7. Pp. 69-128. Oxford University Press, Oxford, England.

Lande, R. 1988. Genetics and demography in biological conservation. Science 241: 1455-1460.

Lesica, P., and F.W. Allendorf. 1995. When are peripheral populations valuable for conservation? Conservation Biology 9: 753-760.

Levin, D.A. 1995. Metapopulations: an arena for local speciation. Journal of Evolutionary Biology 8: 635-644.

McKinney, F.K., S. Lidgard, J.J. Sepkoski, Jr., and P.D. Taylor. 1998. Decoupled temporal patterns of evolution and ecology in two post-Paleozoic clades. Science 281: 807-809.

Moore, W.S., and J.T. Price. 1993. Nature of selection in the Northern Flicker hybrid zone and its implications for speciation theory. *In* Harrison, R.G. (ed.). Hybrid Zones and the Evolutionary Process. Pp. 196-225. Oxford University Press, New York.

Moritz, C. 1994. Defining 'Evolutionarily Significant Units' for conservation. Trends in Ecology and Evolution 9: 373-375.

Reznick, D.N., F.H. Shaw, F.H. Rodd, and R.G. Shaw. 1997. Evaluation of the rate of evolution in natural populations of guppies (*Poecilia reticulata*). Science 275: 1934-1937.

Smith, T.B., R.K. Wayne, D.J. Girman, and M.W. Bruford. 1997. A role for ecotones in generating rainforest biodiversity. Science 276: 1855-1857.

Vane-Wright, R.I., C.L. Lehman, and M.A. Nowak. 1994. Habitat destruction and the extinction debt. Nature 371: 65-66.

Vitousek, P.M., H.A. Mooney, J. Lubchenco, J.M. Melillo. 1997. Human domination of earth's ecosystems. Science 277: 494-499.

GLOSSARY

Allochthonous - Originating outside a designated system. (e.g. From an aquatic point of view, tree leaves falling into a stream are allochthonous because they grew outside the stream.) See *autochthonous*.

Anoxic - Without oxygen.

Autochthonous - Originating inside a designated system. (e.g. From an aquatic point of view, algae growing on the bottom of a stream are autochthonous because they grew inside the stream.) See *allochthonous*.

Basin - See *watershed*.

Benthic - Of or pertaining to the bottom of a river, lake, or other body of water.

Blackwater - A water type generally characterized by low pH, low dissolved oxygen levels, and low nutrient levels. A cup full of this water is dark brown in color, but transparent. Larger bodies of water look black and opaque.

Caño - Literally a canal or a bayou with its own water source. A medium to small stream entering a larger river or tributary; during high water or flooding cycles water will flow into the caño.

Channeling - The dredging and straightening of a river or stream.

Clearwater - A water type generally characterized by low to medium pH, high dissolved oxygen levels, and moderate nutrient levels. The water is transparent.

Cocha - Cochas are oxbow lakes that were former river channels that became isolated during the meander history of the main river. Cochas are connected with their main river during high water but are variable in their connections during low water periods, with either shallow channels or completely disconnected. Some cochas can be isolated from flowing water for months, or perhaps years.

Curiche - See *cocha*.

Dead Arm - A lateral arm of a river where the upriver channel has been closed. There is no or little flow except from tributaries, runoff from rain or during flooding cycles.

Endemic - Found only in a given area, and no where else.

Eutrophication - Large input of nutrients into a body of water which results in increased productivity and decomposition.

Flooded Forest - Flooded forests are broadly open to the rivers and are simply flooded areas within the forest or within large forested islands. Flooded forest areas are only seasonally inundated.

Flooded Lake - A standing lake or body of water over a depressed area, lacking a source or its own drainage basin. Flooded lakes result from the natural flooding cycle; lower or depressed areas holding water after flood waters recede.

Gallery Forest - Forest alongside a river.

Garape - Stream, brook.

Humic - Derived from plant or animal matter being broken down in the soil.

Igarape - Stream, brook.

Lagoon - See *lake*.

Lake - Lakes were taken as different habitats from cochas in that lakes were not obviously oxbows, appeared to have a different formation, and were largely endorheic during low water.

Lentic - Pertaining to still water, as in lakes and ponds. See *lotic*.

Littoral - The aquatic zone extending from the beach to the maximum depth at which light can support the growth of plants.

Lotic - Pertaining to flowing water, as in rivers and streams. See *lentic*.

Macrophytes - A non-microscopic plant.

Miocene - A time period of the earth's history extending from about 25 million years ago to about 5 million years ago.

Ornamental Fishes - Fishes considered appropriate for the aquarium business.

Oxbow Lake - See *cocha*.

Periphyton - Algae attached to rocks, logs, and other underwater substrates.

Permian - A time period of the earth's history extending from about 290 million years ago to about 260 million years ago.

Planktonic - Free floating or drifting in the water; not swimming through the water or attached to a structure.

Precambrian - A time period of the earth's history extending from the origin of the earth to about 580 million years ago.

Quaternary - A time period of the earth's history extending from about 2 million years ago to the present.

Río - River.

Riparian - Found along the edge of a river. Often used in the context of vegetation.

Seine - A mesh net, often used to catch fish and other larger aquatic organisms.

Semi-lotic - Pertaining to water flowing slowly. Intermediate between lentic and lotic. See *lentic* and *lotic*.

Terra Firme - High ground beyond the flood plain of a river. See *varzea*.

Tertiary - A time period of the earth's history extending from about 65 million years ago to about 2 million years ago.

Varzea - Flood plain of a river. See *terra firme*.

Watershed - A region drained by a particular river and its associated streams. Also known as a basin.

Whitewater - A water type generally characterized by medium pH, moderate dissolved oxygen levels, and high nutrient levels. Although it is called whitewater, it is actually brown (café au lait) in color due to high turbidity.

APPENDICES

APPENDIX 1 Summary of the physico-chemical characteristics of the waters in the Upper Tahuamanu and Lower Nareuda/Middle Tahuamanu sub-basins during the AquaRAP expedition to Pando, Bolivia in September 1996.

Francisco Antonio R. Barbosa, Fernando Villarte V., Juan Fernando Guerra Serrudo, Germana de Paula Castro Prates Renault, Paulina María Maia-Barbosa, Rosa María Menéndez, Marcos Callisto Faria Pereira, and Juliana de Abreu Vianna

UPPER TAHUAMANU SUB-BASIN

Site 1: Río Tahuamanu, left margin
(11° 26.699'S; 69° 0.509'W)
Date: 4/9/96 ; Time: 9:00 am
Depth: 0.25m; three hauls for benthos; substrate: fine sand.
Water temp.: 28.4°C; pH: 7.43; cond.: 270 µS/cm; oxyg.: 8.85 mg/l, 114.3% sat.
Left margin: sand beach
Site 1: Right Margin
Time: 10:30 am
Depth:1.5 m; three hauls for benthos; substrate: fine sand.
Water temp.: 28.8°C; cond.: 240 µS/cm; pH: 7.50; oxyg.: 9.29 mg/l, 120.5% sat.
Right margin: riparian forest

Site 2. Lake at Aceradero
(11° 25.802'S; 69° 0.481'W)
Date: 4/9/96; Time: 1:40 pm
Depth: 0.5 m; distance from shore: 1.5 m, near a stand of macrophytes (*Pistia* sp and *Scirpus* sp); substrate: mud.
Water temp.: 30.8°C; cond.: 27.4 µS/cm; pH: 6.40; oxyg.: 9.45 mg/l, 124.1 % sat.
Site 2. Outlet of the lake (11° 25.872'S; 69° 0.189'W)
Water temp.: 29.1°C; cond.: 30.9 µS/cm; pH: 6.46; oxyg.: 10.0 mg/l, 126.7 % sat.

Site 3: Río Muymano, left margin
(11° 27.225'S; 69° 01.810'W)
Date: 5/9/96; Time: 9:50 am
Depth: 1.30 m; distance from shore: approx. 3 m
Water temp.: 27.8°C; cond.: 78.9 µS/cm; pH: 6.85; oxyg.: 9.47 mg/l, 124.2 % sat.

Site 4: Lake/Río Muymano
(11° 27.197'S; 69° 1.834'W)
Date: 5/9/96; Time: 10:25 am
Depth: 0.50m; distance from shore: approx. 1 m; substrate: mud
Water temp.: 28.7°C; cond.: 122.8 µS/cm; pH: 6.72; oxyg.: 8.47 mg/l, 111.0 % sat.

Site 5: Río Tahuamanu, left margin
(11° 26.792'S; 69° 1.434'W)
before the mouth of the Muymano river.
Date: 5/9/96; Time: 12:25
Depth: 0.50 m; distance from shore: approx. 1 m; sand beach.
Water temp.: 31.2°C; cond.: 355.0 µS/cm; pH: 7.59; oxyg.: 10.69 mg/l, 154.6 % sat.

Site 6: Mouth of Río Muhymano, right margin
(11° 26.839'S; 69° 01.468'W)
Date: 5/9/96; Time: 1:30 pm
Depth: > 2.0 m; distance from shore: approx. 3.0 m; riparian vegetation.
Water temp.: 28.3°C; cond.: 83.0 µS/cm; pH: 6.97; oxyg.: 10.41 mg/l, 134.6 % sat.

Site 7: Lake at right margin Río Tahuamanu
(11° 26.723'S; 69° 0.692'W)
Date: 7/9/96; Time: 10:00 am; Weather: cloudy
Depth: 1.10 m; distance from shore: approx. 3.0 m; substrate: mud
Water temp.: 26.2°C; cond.: 59.9 µS/cm; pH: 6.83; oxyg.: 8.82 mg/l, 108.4 % sat.

Site 8: Lake Remanso, left margin Río Tahuamanu (11° 26.054'S; 69° 0.879'W)
Date: 7/9/96; Time: 2:00 pm
Depth: 1.0 m; distance from shore: approx. 3.0 m; substrate: mud
Water temp.: 27.5°C; cond.: 92.7 µS/cm; pH: 6.66; oxyg.: ?

Site 9: Lake Canhaveral, left margin Río Tahuamanu (11° 26.257'S; 69° 1.997'W)
Date: 7/9/96; Time: 4:30 pm; substrate: ; abundant Neuston
Depth: 0.50 m; distance from shore: approx. 2.0 m
Water temp.: 31.0°C; cond.: 92.4 µS/cm; pH: 7.90; oxyg.: 0.79 mg/l, 10.3% sat.

Site 10: Curiche at left margin Río Tahuamanu (11° 25.263'S; 69° 5.160'W)
Date: 8/9/96; Time: 12:30
Remnant of the river entirely covered by vegetation; dark water, probably containing humic substances and ferrous bacteria.
Depth: 0.70m; distance from shore: 0.30 m
Water temp.: 22.2°C; cond.: 31.0 µS/cm; pH: 5.46; oxyg.: 2.21 mg/l, 23.0% sat.

LOWER NAREUDA/MIDDLE TAHUAMANU SUB-BASIN

Site 11: Curiche at right margin Río Nareuda (11° 18.336'S; 68° 45.861'W)
Date: 10/9/96; Time: 11:00 am
Depth: 0.50 m; distance from shore: 1.0 m; substrate: clay
Water temp.: 22.8°C; cond.: 40.2 µS/cm; pH: 6.05; oxyg.: 3.48 mg/l, 39.5% sat.

Site 12: Río Nareuda, right margin (11° 18.541'S; 68° 45.938'W)
Date: 10/9/96; Time: 12:15
Depth: 1.5 m; distance from shore: approx. 3.0 m; substrate: fine sand
River width: approx. 15 m; sampling area for benthos: approx. 5 m^2;
Water temp.: 24.1°C; cond.: 44.2 µS/cm; pH: 6.71; oxyg.:7.37 mg/l, 84.3% sat.

Site 13: Curiche at right margin of Río Nareuda (11° 18.525'S; 68° 45.965'W)
Date: 11/9/96; Time: 10:00 am
Depth: 0.50 m; distance from shore: approx. 1.0 m; substrate: clay
Water temp.: 19.7°C; cond.: 40.6 µS/cm; pH: 6.00; oxyg.: 3.08 mg/l, 31.6% sat.

Site 14: Río Nareuda, right margin, at the rapids (11° 18.306'S; 68° 45.422'W)
Date: 11/9/96; Time: 01:20 pm
Depth: 0.50 m; distance from shore: approx. 3.0 m; substrate: stones/gravel
River width: approx. 15 m; At this site it was collected periphyton from the stones.
Water temp.: 23.9°C; cond.: 51.4 µS/cm; pH: 6.70; oxyg.: 7.91 mg/l, 94.8% sat.

Site 15: Arroyo Filadelfia, right margin Río Tahuamanu (11° 20.550'S; 68° 45.895'W)
Date: 12/9/96; Time: 11:00 am
Depth: 0.80 m; distance from shore: approx. 1.0 m; stream width: approx. 3.0 m
substrate: clay;
Water temp.: 21.1°C; cond.: 12.3 µS/cm; pH: 5.86; oxyg.: 7.83 mg/l, 86.1% sat.

Site 16: Río Tahuamanu, right margin (11° 20.567'S; 68° 46.061'W)
Below the mouth of Arroyo Fila
Date: 12/9/96; Time: 01:30 pm
Depth: > 2.0 m; distance from shore: approx. 3.0 m
Water temp.: 26.7°C; cond.: 175.0 µS/cm; pH: 7.46; oxyg.: 9.31 mg/l, 116.6% sat.

Site 17: Lake Filadelfia, right margin Río Tahuamanu (11° 20.026'S; 68° 45.353'W)
Date: 12/9/96; Time: 02:30 pm
Depth: 1.0 m; distance from shore: approx. 3.0 m; substrate: clay
Water temp.: 28.6°C; cond.: 102.6 µS/cm; pH: 7.19; oxyg.: 8.65 mg/l, 112.0% sat.

Site 18: Lake at right margin Río Tahuamanu (11° 18.61'S; 68° 44.425'W)
Date: 12/9/96; Time: 05:25 pm
Depth: 1.0 m; distance from shore: approx. 2.0 m; substrate: clay
Water temp.: 26.2°C; cond.: 22.1 µS/cm; pH: 5,71; oxyg.: 2.49 mg/l, 30.7% sat.

Site 19: Igarapé Preto, left margin Río Tahuamanu (11° 16.273'S; 68° 44.338'W)
Date: 13/9/96; Time: 10:20 am
Depth: 0.70 m; distance from shore: approx. 2.0m; substrate: clay
Black water; sampling area for benthos: approx. 10 m2.
Water temp.: 22.5°C; cond.: 17.5 µS/cm; pH: 6.2; oxyg.: 8.05 mg/l, 87.0% sat.

Physico-chemical characteristics of the waters in the Upper Tahuamanu and Lower Nareuda/Middle Tahuamanu sub-basins

Site 20: Lake at right margin of Río Tahuamanu (11° 17.609'S; 68° 44.494'W)
The lake is permanently conected to the river; clear water
Date: 13/9/96; Time: 11:30 am
Depth: 1.0 m; distance from shore: approx. 1.5 m; substrate: clay
Water temp.: 28.2°C; cond.: 331.0 µ S/cm; pH: 7.62; oxyg.: 8.34 mg/l, 107.3% sat.

Site 21: Río Tahuamanu, central channel at the rapids (11° 18.273'S; 68° 44.482'W)
Date: 13/9/96; Time: 12:30;
Depth: 0.80 m; distance from shore: approx. 40 m; substrate: rocks
Clear water; periphyton present on the rocks.
Water temp.: 27.4°C; cond.: 159.2µS/cm; pH: 7.38; oxyg.: 7.51 mg/l, 88.0% sat.

Site 22: Arroyo las Alicias, right margin Río Tahuamanu (11° 18.493'S; 68° 44.585'W)
Date: 13/9/96; Time: 01:00 pm
Depth: 0.50 m; distance from shore: 1.0 m; substrate: clay/ stones
Water color: white/dark
Water temp.: 25.0°C; cond.: 13.7 µS/cm; pH: 6.40; oxyg.: 7.48 mg/l, 88.3% sat.

Distribution of zooplankton species along the sampling stations of the AquaRAP expedition to Pando, Bolivia in September 1996.

APPENDIX 2

Francisco Antonio R. Barbosa, Fernando Villarte V., Juan Fernando Guerra Serrudo, Germana de Paula Castro Prates Renault, Paulina María Maia-Barbosa, Rosa María Menéndez, Marcos Callisto Faria Pereira, and Juliana de Abreu Vianna

+ = present; - = absent

RECORDED TAXA	SAMPLING STATIONS (see Appendix 1)											
	1 S.E.	1 N.W.	2	3	4	5	6	7 limn.	7 litt.	8	9	10
PROTOZOA - RHIZOPODA												
Arcellidae												
Arcella costata	-	-	+	-	-	-	-	-	-	-	-	-
A. discoides	+	+	-	+	-	-	+	-	-	+	-	-
A. discoides v. pseudovulgaris	-	-	+	-	-	-	-	-	-	-	-	-
A. gibbosa	-	-	-	-	-	-	-	-	-	+	-	-
A. gibbosa v. mitriformes	-	-	+	+	-		-	-	-	-	-	-
A. hemisphaerica	+	-	-	+	+	+	-	-	-	+	-	-
A. hemisphaerica undulata	-	+	+	+	+	-	+	+	+	+	-	-
A. megastoma	-	-	-	-	-	-	-	-	-	+	-	-
A. mitrata	+	-	-	-	-	-	-	-	-	-	-	-
A. rotundata	+	+	-	+	+	-	-	-	-	-	-	-
A. vulgaris	-	-	-	-	+	-	-	-	-	+	-	+
A. vulgaris v. undulata	-	-	-	-	-	-	-	-	-	+	-	-
Centropyxidae												
Centropyxis sp.	-	+	-	-	-	+	-	-	-	-	-	-
C. aculeata	-	+	+	+	-	-	+	-	-	-	-	-
C. ecornis	+	-	-	+	-	-	-	-	-	+	-	-
C. marsupiformes	-	+	-	-	-	-	-	-	-	-	-	-
C. platystoma	-	-	-	-	-	-	+	-	-	-	-	-
Cyclopyxis arcelloides	-	-	-	-	-	-	+	-	-	-	-	-
C. euristoma	-	-	+	-	-	-	-	-	-	-	-	-
Diffugidae												
Difflugia acuminata	+	-	-	-	+	-	-	-	-	-	-	-
D. acutissima	+	-	-	-	+	-	-	-	-	-	-	-
D. avellana	-	-	-	+	+	-	-	-	-	-	-	-
D. corona	-	-	-	-	-	-	-	-	-	+	-	-
D. elegans	-	-	-	-	+	-	-	-	-	-	-	-
D. gramen	+	+	+	+	+	-	-	-	-	-	-	-
D. oblonga	-	-	-	-	-	-	-	-	-	-	-	+
D. oblonga v. cylindrus	-	-	-	-	+	-	-	-	-	-	-	-
D. lanceolata	+	-	-	-	-	-	+	-	-	-	-	-
D. lobostoma	-	-	+	-	-	-	-	-	-	-	-	-
D. lobostoma fma multilobada	-	+	+	-	-	-	-	-	-	-	-	-
D. lithophila	-	+	+	-	-	-	-	-	-	-	-	-
D. difficilis	-	+	-	-	-	+	-	-	-	-	-	-
D. smilion?	-	+	-	-	-	-	-	-	-	-	-	-
Cucurbitella sp.	-	-	-	-	-	-	-	+	-	-	-	-
C. mespiliformes	-	-	-	-	-	-	-	-	+	+	-	-
C. obturata	-	-	-	-	+	-	-	-	-	-	-	-
Protocucurbitella coroniformes	-	-	+	-	-	-	-	-	-	-	-	-
P. microstoma	-	-	-	-	-	-	-	-	+	-	-	-

RECORDED TAXA	SAMPLING STATIONS (see Appendix 1)											
	1 S.E.	1 N.W.	2	3	4	5	6	7 limn.	7 litt.	8	9	10
Euglyphidae												
Euglypha sp.	-	-	-	-	+	-	+	-	-	-	-	-
E. acanthophora	+	-	+	+	-	-	-	-	-	+	-	-
E. acanthophora v. cylindracea	-	-	-	-	-	+	-	-	-	-	-	-
E. tuberculata	+	+	-	+	-	-	-	-	-	-	-	+
Eughypha laevis	-	+	+	-	-	-	+	-	-	-	-	-
Trinema lineare	+	+	+	-	-	-	+	-	-	-	-	-
T. enchelys	+	-	-	+	-	-	-	-	-	+	-	-
Nebellidae												
Lesquereusia modesta	-	-	+	-	-	-	-	-	-	-	-	-
Nebela sp.	-	-	-	-	-	-	-	-	-	-	-	+
Quadrulella symetrica	-	-	-	-	-	-	+	-	-	-	-	-
Sub-total 1	13	14	15	12	12	4	10	2	3	12	0	4
ROTIFERA												
Brachionidae												
Anuraeopsis fissa	+	-	-	-	-	+	+	+	+	+	+	-
Brachionus caudatus	+	+	+	-	-	+	-	+	+	+	+	-
Brachionus dolabratus	-	-	-	-	-	-	-	+	-	-	-	-
Brachionus falcatus falcatus	+	-	-	-	+	-	-	+	-	-	-	-
Brachionus mirus	-	-	-	-	-	-	-	+	+	-	-	-
Brachionus urceolaris	+	+	-	-	-	-	-	-	-	+	-	-
Epiphanes sp.	+	-	-	-	-	-	-	-	-	-	-	-
Keratella cochlearis	-	-	+	-	-	-	-	+	-	-	+	-
K. tropica	+	+	+	+	+	-	-	+	-	+	-	-
Platyias quadricornis	-	-	-	-	-	-	-	-	+	-	-	+
Plationus patulus	-	-	-	-	-	-	-	-	-	+	-	-
Euchlanidae												
Euchlanis dilatata	-	-	-	+	+	-	-	-	-	-	-	-
Notommatidae								-				
Cephalodella gibba	+	+	-	+	-	+	-	+	-	-	-	-
Cephalodella gigantea	-	-	-	-	-	-	+	-	-	-	-	-
Cephalodella sp. 1	-	-	-	-	+	-	-	-	-	+	-	-
Cephalodella sp. 2	-	-	-	-	+	-	+	-	-	-	-	-
Notommata sp.	+	-	-	-	-	+	-	-	+	-	-	-
Scaridium longicaudum	-	-	-	-	-	-	-	-	-	+	-	-
Colurellidae												
Colurella sp.	-	-	+	-	-	-	-	-	-	-	-	+
C. obtusa	+	-	-	+	-	-	-	-	+	-	-	+
Lepadella sp.	-	-	-	-	-	-	-	-	+	-	-	-
L. patella	-	-	-	+	-	-	+	-	-	+	-	-
L. rhomboides	-	-	+	-	-	-	-	-	-	-	-	-
L. latusinus	-	-	-	+	-	-	-	-	-	-	-	-

RECORDED TAXA	SAMPLING STATIONS (see Appendix 1)											
	1 S.E.	1 N.W.	2	3	4	5	6	7 limn.	7 litt.	8	9	10
Mytilinidae												
Mytilina ventralis	+	-	-	-	-	-	-	-	-	-	-	-
Filinidae												
Filinia longiseta	-	-	+	-	-	-	-	+	+	+	+	-
Hexarthridae												
Hexarthra sp.	-	-	+	-	-	-	-	+	+	-	+	-
Philodinidae												
Dissotrocha macrostyla	-	-	-	+	-	-	-	-	-	-	-	-
Rotaria rotatoria	+	-	+	-	-	-		-	+	-	-	-
Bdelloidea	+	-	+	+	+	+	-	-	-	+	-	+
Lecanidae												
Lecane bulla	-	+	+	-	-	-	-	-	-	+	-	-
L. closterocerca	-	-	-	-	-	-	+	-	-	-	-	-
L. furcata	-	-	+	-	-	-	-	-	-	-	-	-
L. hamata	-	-	-	-	-	-	-	-	-	+	-	-
L. hastata	+	+	-	-	-	-	-	-	-	-	-	-
L. inermis	-	-	-	-	-	-	+	-	-	-	-	-
L. leontina	-	-	+	-	-	-	-	-	-	-	-	-
L. lunaris	-	-	-	-	-	-	-	-	-	+	-	-
L. papuana	+	+	-	-	+	+	-	-	-	-	+	-
L. pyriformes	-	+	-	-	-	-	-	-	-	-	-	-
L. stichaea	-	-	-	+	-	-	-	-	-	-	-	-
Synchaetidae												
Polyarthra sp.	-	-	+	+	+	-	-	+	+	+	+	-
P. dolychoptera	-	-	+	-	-	-	-	-	-	-	-	-
Synchaeta sp.	-	-	+	-	-	-	-	-	-	-	-	-
Testudinellidae												
Pompholyx sulcata	-	+	-	-	-	+	-	-	-	+	-	+
Testudinella patina patina	-	+	+	-	-	-	-	-	-	+	-	-
Trichocercidae												
Trichocerca (D) porcellus	-	-	+	+	+	-	-	-	-	-	-	-
T. iernis	-	-	-	-	+	-	-	-	-	+	-	-
T. insignis	+	-	+	-	-	-	-	-	-	-	-	-
T. pussila	+	-	-	-	-	-	+	+	+	+	+	-
T. rousseletie?	+	-	-	-	-	-	-	-	-	-	-	-
T. similis	+	-	-	-	-	-	-	-	-	-	+	+
T. tenuior	-	+	-	-	-	-	-	-	-	-	-	-
Sub-total 2	18	11	18	11	10	7	7	12	12	18	9	6

RECORDED TAXA	SAMPLING STATIONS (see Appendix 1)											
	1 S.E.	**1 N.W.**	**2**	**3**	**4**	**5**	**6**	**7 limn.**	**7 litt.**	**8**	**9**	**10**
COPEPODA												
Nauplius	+	-	+	+	+	+	-	+	+	-	+	+
Copepodito Cyclopoida	+	-	+	+	+	-	-	-	+	-	-	+
Cryptocyclops brevifurca	-	-	+	-	-	-	-	-	-	-	-	-
Mesocyclops sp.	-	-	+	-	-	-	-	-	-	-	-	-
Microcyclops sp.	-	-	+	-	-	-	-	-	-	-	-	-
Tropocyclops prasinus	+	-	-	-	-	-	-	-	-	-	-	-
Notodiaptomus sp.	-	-	-	+	-	-	-	-	-	-	-	-
Copepodito Calanoida	-	-	-	+	-	+	-	-	-	-	-	-
Calanoida	-	-	-	-	+	-	-	-	-	-	-	-
Ergasilus sp.	-	-	-	-	-	-	-	-	+	-	+	-
Sub-total 3	1	0	3	1	1	0	0	0	1	0	1	0
CLADOCERA												
Ilyocryptidae												
Ilyocryptus spinifer	+	+	+	-	-	+	-	-	-	+	-	-
Moinidae												
Moina minuta	+	-	+	+	+	+	-	-	-	-	+	-
Dapnidae												
Ceriodaphnia cornuta	-	+	-	-	-	-	-	-	-	-	-	-
Moinodaphnia macleayi	-	-	+	-	-	-	-	-	-	-	-	-
Simocephalus serrulatus	-	+	-	-	-	-	-	-	-	-	-	-
Macrothricidae												
Macrothrix laticornis	+	+	-	-	-	+	-	-	-	-	-	-
Chydoridae												
Alonella hamulata	-	-	+	-	-	-	-	-	-	-	-	-
Eurialona orientalis	-	-	+	-	-	-	-	-	-	-	-	-
Oxyurella longicaudis	-	-	+	+	-	-	-	-	-	-	-	-
Sub-total 4	3	4	6	2	1	3	0	0	0	1	1	0
OSTRACODA	-	-	+	+	+	-	-	-	+	-	-	-
Sub-total 5	0	0	1	1	1	0	0	0	1	0	0	0
GASTROTRICHA												
Chaetonotus sp.	-	-	+	-	-	+	-	+	+	+	-	+
Sub-total 6	0	0	1	0	0	1	0	1	1	1	0	1
NEMATODA	+	+	+	-	+	+	+	-	-	-	-	+
Sub-total 7	1	1	1	0	1	1	1	0	0	0	0	1
TOTAL	**36**	**30**	**45**	**27**	**26**	**16**	**18**	**15**	**18**	**32**	**11**	**12**

Composition of the benthic macroinvertebrate community in selected sites of the AquaRAP expedition to Pando, Bolivia in September 1996.

APPENDIX 3

Francisco Antonio R. Barbosa, Fernando Villarte V., Juan Fernando Guerra Serrudo, Germana de Paula Castro Prates Renault, Paulina María Maia-Barbosa, Rosa María Menéndez, Marcos Callisto Faria Pereira, and Juliana de Abreu Vianna

	SAMPLING SITES (See Appendix 1)							
TAXA	**SITE 1 - N.W.**	**SITE 2**	**SITE 8**	**SITE 9**	**SITE 13**	**SITE 16**	**SITE 18**	**SITE 20**
Mollusca								
Bivalvia	44	1	181	77	10	75	18	48
Gastropoda *Biomphalaria*	-	-	-	-	-	-	1	-
Oligochaeta	6	6	26	3	30	79	14	3
Hirudinea	1	-	2	1	-	13	-	2
Decapoda - Gamaridae	-	-	2	1	-	2	-	-
Acarina	-	-	-	-	-	-	-	-
Arachnida	-	-	-	-	-	-	1	-
Insecta								
Odonata	1	2	18	5	9	3	12	5
Heteroptera	2	-	106	7	22	155	7	2
Lepidoptera	-	-	-	-	-	-	1	-
Ephemeroptera	2	-	69	7	-	17	2	18
Trichoptera	1	1	7	-	1	-	-	3
Coleoptera	3	3	-	-	9	9	2	1
Diptera								
Ceratopogonidae	9	1	4	11	17	50	7	2
Chaoboridae	-	-	-	1	-	-	3	1
Chironomidae	27	11	107	22	32	51	25	120
Culicidae	-	-	1	-	-	-	-	-
Tabanidae	-	-	-	-	1	-	-	-
Tipulidae	-	3	-	-	-	-	-	-
Total Number of Individuals	96	28	523	135	131	454	93	205

APPENDIX 4

Distribution of Chironomidae genera identified from selected sites of the AquaRAP expedition to Pando, Bolivia in September 1996.

Francisco Antonio R. Barbosa, Fernando Villarte V., Juan Fernando Guerra Serrudo, Marcos Callisto Faria Pereira, and Juliana de Abreu Vianna

	SAMPLING SITES (See Appendix 1)							
TAXA	SITE 1 - N.W.	SITE 2	SITE 8	SITE 9	SITE 13	SITE 16	SITE 18	SITE 20
TANYPODINAE								
Ablabesmyia	-	-	18	1	-	3	1	2
Coelotanypus	14	-	18	15	-	27	2	26
Djalmabatista	-	-	-	-	-	-	1	-
Labrundinia	-	-	2	-	2	1	2	2
CHIRONOMINAE								
Asheum	-	-	10	-	1	-	-	6
Chironomus	-	2	40	-	-	-	5	3
Cladopelma	-	-	-	-	-	-	1	-
Fissimentum desiccatum	-	-	-	-	-	9	5	12
Goeldichironomus	2	-	2	1	2	1	-	48
Harnischia	-	1	1	-	-	-	-	2
Parachironomus	-	-	1	-	1	-	-	-
Polypedilum	10	-	2	5	2	4	5	8
Stenochironomus	-	-	6	-	-	-	-	-
Tribelos	1	1	1	-	-	-	-	-
Zavreliella	-	-	1	-	-	-	-	-
Nimbocera paulensis	-	5	-	-	10	-	-	4
Tanytarsini genera	-	2	-	1	14	5	3	5
Richness	4	5	12	5	7	7	9	11
Total Number of Individuals	27	11	102	23	32	50	25	118
H' Shannon-Wiever (base 2)	1.48	2.04	2.58	1.47	2.11	2.02	2.90	2.62
Equitabilidade	0.74	0.88	0.72	0.63	0.75	0.72	0.92	0.76

Decapod crustaceans collected by the AquaRAP expedition to Pando, Bolivia in September 1996.

Célio Magalhães

Field Station	Number of individuals collected	Species
B96 – P01 – 01	1	*Palaemonetes ivonicus* (imat.)
B96 – P01 – 02	10	*Macrobrachium depressimanum*
B96 – P01 - 03	1 m	*Sylviocarcinus devillei*
	39	*Macrobrachium depressimanus*
B96 – P01 –04	1	*Macrobrachium depressimanum*
B96 – P01 – 08	7	*Macrobrachium depressimanum*
	4	*Palaemonetes ivonicus* (juv.)
B96 – P01 – 09	1 f	*Valdivia* cf. *serrata* (imat.)
B96 – P01 – 10	1	*Palaemonetes ivonicus*
	1	*Macrobrachium jelskii*
	1	*Macrobrachium amazonicum*
B96 - P01 - 11	2	*Macrobrachium* cf. *brasiliense* (imat.)
	10	*Macrobrachium jelskii*
	17	*Macrobrachium depressimanum*
	1 m	*Sylviocarcinus maldonadoensis*
	2 m	*Zilchiopsis oronensis*
B96 - P01 - 12	2	*Macrobrachium* cf. *brasiliense*
	20	*Palaemonetes ivonicus*
B96 - P01 - 13	1	*Macrobrachium depressimanum*
	1	*Macrobrachium brasiliense*
	9	*Palaemonetes ivonicus* (imat. + juv.)
B96 - P01 - 14	1	*Macrobrachium depressimanum*
	1	*Macrobrachium brasiliense*
	3	*Macrobrachium jelskii*
B96 - P01 - 14	29	*Palaemonetes ivonicus*
B96 - P01 - 17	16	*Macrobrachium jelskii*
B96 - P01 - 18	1	*Macrobrachium depressimanum*
	1 f	*Sylviocarcinus maldonadoensis*
B96 - P01 - 20	1	*Acetes paraguayensis*
B96 - P01 - 23	2	*Palaemonetes ivonicus*
B96 - P01 - 25	1	*Palaemonetes ivonicus*
B96 - P01 - 26	1	*Macrobrachium amazonicum*
	2	*Palaemonetes ivonicus*
	3 f	*Valdivia* cf. *serrata* (imat.)
B96 - P01 - 30	1	*Macrobrachium amazonicum*
B96 - P01 - 31	1	*Acetes paraguayensis*
	1	*Macrobrachium jelskii*
	4	*Macrobrachim depressimanus*
B96 - P01 - 32	5	*Macrobrachium depressimanum*
B96 - P01 - 33	3	*Macrobrachium amazonicum*

P01 = Group consisted of Jaime Sarmiento, Hernán Ortega, Soraya Barrera, and Luis Fernando Yapur.
F = Female; M= Male

Field Station	Number	Species
B96 - P02 – 01	28	*Macrobrachium depressimanum*
	17	*Macrobrachium brasiliense*
	2 m	*Zilchiopsis oronensis*
	2 m	*Valdivia* cf. *serrata*
B96 - P02 – 02	5	*Macrobrachium depressimanum*
	10	*Macrobrachium jelskii*
	2 m	*Valdivia* cf. *serrata*
B96 - P02 – 03	2	*Macrobrachium depressimanum*
B96 - P02 – 05	16	*Macrobrachium depressimanum*
B96 - P02 – 06	15	*Macrobrachium brasiliense*
B96 - P02 – 07	1	*Macrobrachium* cf. *brasiliense*
	2	*Macrobrachium jelskii*
	3	*Macrobrachium depressimanum*
B96 - P02 – 08	4	*Macrobrachium depressimanum*
B96 - P02 – 09	1 f	*Valdivia* cf. *serrata* (imat.)
B96 - P02 – 12	2	*Macrobrachium* cf. *brasiliense*
B96 - P02 – 13	1	*Macrobrachium* cf. *brasiliense*
B96 - P02 – 14	6	*Acetes paraguayensis*
	10	*Macrobrachium depressimanum*
B96 - P02 – 15	2	*Macrobrachium depressimanum*
B96 - P02 – 16	3	*Acetes paraguayensis*
	4	*Macrobrachium depressimanum*
B96 - P02 – 17	2	*Macrobrachium depressimanum*
B96 - P02 – 18	3	*Macrobrachium* cf. *brasiliense*
B96 - P02 – 22	4	*Macrobrachium depressimanum*
	1 f	*Sylviocarcinus devillei* (imat.)
B96 - P02 – 23	2	*Macrobrachium depressimanum*
B96 - P02 – 26	6	*Macrobrachium depressimanum*
B96 - P02 – 27	34	*Acetes paraguayensis*
	17	*Macrobrachium depressimanum*
	1 f	*Sylviocarcinus maldonadoensis*
B96 - P02 – 32	4	*Acetes paraguayensis*
B96 - P02 – 34	1	*Macrobrachium amazonicum*
B96 - P02 – 36	3	*Macrobrachium amazonicum*
	39	*Macrobrachium depressimanum*
B96 - P02 – 43	70	*Macrobrachium amazonicum*
	15	*Macrobrachium amazonicum*
	1	*Acetes paraguayensis*

P02 = Group consisted of Barry Chernoff, Naércio Menezes, Theresa M. Bert, Roxana Coca, and Antonio Machado-Allison.
F = Female; M = Male

Fishes collected during the AquaRAP expedition to Pando, Bolivia in September 1996.

APPENDIX 6

Jaime Sarmiento, Barry Chernoff, Soraya Barrera, Antonio Machado-Allison, Naércio Menezes, and Hernán Ortega

TAXA	Upper Nareuda	Lower Nareuda	Upper Tahuamanu	Middle Tahuamanu	Lower Tahuamanu	Manuripi
RAJIFORMES						
Potamotrygonidae						
Potamotrygon motoro	-	●	●	-	-	-
CLUPEIFORMES						
Engraulidae						
Anchoviella cf. *carrikeri*	-	-	-	●	-	●
CHARACIFORMES						
Anostomidae						
Abramites hypselonotus	-	●	-	●	-	●
Laemolyta sp.	-	-	-	-	-	●
Leporinus cf. *fasciatus*	-	-	-	-	-	●
Leporinus friderici	●	-	●	-	-	●
Leporinus cf. *nattereri*	-	●	-	-	-	●
Schizodon fasciatum	●	-	●	-	-	●
Characidae						
Aphyocharax alburnus	●	●	-	-	-	-
Aphyocharax dentatus	●	-	●	●	●	●
Aphyocharax pusillus	-	-	●	●	-	●
Astyanax cf. *abramis*	●	●	●	●	●	●
Astyanax sp.	-	-	●	-	-	-
Brachychalcinus copei	●	●	●	-	-	-
Bryconamericus cf. *caucanus*	●	-	-	-	-	-
Bryconamericus cf. *pachacuti*	●	-	-	-	-	-
Bryconamericus cf. *peruanus*	●	●	-	-	-	-
Bryconamericus sp.	●	-	-	●	-	-
Characidium sp. 1	●	●	●	●	-	-
Characidium sp. 2	●	-	-	●	-	-
Charax gibbosus	-	●	-	-	-	-
Cheirodon fugitiva	-	●	●	●	-	●
Cheirodon sp. 1	-	-	-	-	-	●
Cheirodon sp. 2	-	●	-	-	-	-
Cheirodontinae sp.	-	-	-	-	-	●
Chrysobrycon sp. 1	●	-	-	-	-	-
Chrysobrycon sp. 2	●	-	●	-	-	-
Clupeacharax anchoveoides	-	-	●	●	-	-
Creagrutus sp. 1	●	-	-	-	-	-
Creagrutus sp. 2	●	●	●	●	-	●
Creagrutus sp. 3	●	-	-	●	-	-
Ctenobrycon spilurus	-	●	●	●	●	●
Cynopotamus gouldingi	●	●	-	-	-	-
Engraulisoma taeniatum	-	-	●	●	●	●

TAXA	Upper Nareuda	Lower Nareuda	Upper Tahuamanu	Middle Tahuamanu	Lower Tahuamanu	Manuripi
Eucynopotamus biserialis	-	-	-	●	●	●
Galeocharax gulo	●	-	●	●	●	-
Gephyrocharax sp.	●	-	●	-	-	-
Hemigrammus lunatus	-	-	-	●	-	●
Hemigrammus cf. *megaceps*	-	-	-	●	-	-
Hemigrammus ocellifer	●	●	-	●	-	●
Hemigrammus cf. *pretoensis*	-	-	-	-	-	●
Hemigrammus? sp.	-	-	-	-	-	●
Hemigrammus cf. *unilineatus*	-	-	-	-	-	●
Hyphessobrycon agulha	●	-	-	-	-	-
Hyphessobrycon cf. *anisitsi*	-	-	-	-	-	●
Hyphessobrycon cf. *gracilior*	●	-	-	●	-	●
Hyphessobrycon? sp.	-	-	-	-	-	●
Hyphessobrycon cf. *tucunai*	-	-	-	●	-	-
Iguanodectes spilurus	-	-	-	-	-	
Knodus cf. *caquetae*	-	-	-	●	-	●
Knodus cf. *gamma*	●	●	●	●	-	●
Knodus cf. *heterestes*	-	-	-	●	-	●
Knodus sp.	●	●	-	●	●	●
Knodus cf. *victoriae*	●	●	●	●	●	●
Metynnis luna	-	-	-	-	-	●
Microschemobrycon geisleri	●	-	-	●	-	-
Moenkhausia cf. *chrysargyrea*	-	-	-	-	-	●
Moenkhausia colletti	●	●	●	●	-	●
Moenkhausia cf. *comma*	-	-	-	-	-	●
Moenkhausia dichroura	●	●	●	●	●	●
Moenkhausia cf. *jamesi*	-	-	-	-	●	●
Moenkhausia cf. *lepidura*	-	●	-	●	●	●
Moenkhausia cf. *megalops*	-	-	-	-	-	●
Moenkhausia sanctaefilomenae	●	●	●	●	-	●
Moenkhausia sp. 1	-	●	-	-	-	-
Moenkhausia sp. 2	-	●	-	-	-	●
Moenkhausia sp. 3	●	●	●	●	●	●
Moenkhausia sp. 4	-	-	-	●	-	-
Moenkhausia sp. 5	-	-	-	-	-	●
Moenkhausia sp. 6	-	-	-	-	-	●
Moenkhausia sp. 7	-	-	-	-	-	●
Moenkhausia sp. 8	-	-	-	-	-	●
Myleus sp.	-	●	-	-	-	-
Mylossoma duriventris	●	-	●	-	-	●

TAXA	Upper Nareuda	Lower Nareuda	Upper Tahuamanu	Middle Tahuamanu	Lower Tahuamanu	Manuripi
Odontostilbe hasemani	●	-	●	●	-	-
Odontostilbe piaba	-	-	-	-	-	●
Odontostilbe paraguayensis	●	●	●	●	-	●
Odontostilbe sp. 1	-	●	●	-	-	-
Odontostilbe sp. 2	-	●	-	-	-	●
Paragoniates alburnus	●	●	●	●	●	●
Phenacogaster cf. *microstictus*	-	●	-	-	-	●
Phenacogaster cf. *pectinatus*	●	●	●	●	-	●
Phenacogaster sp. 1	●	-	●	●	-	●
Phenacogaster sp. 2	●	●	●	-	●	●
Phenacogaster sp. 3	-	-	-	●	-	●
Phenacogaster? sp.	●	-	●	●	●	-
Piabucus melanostomus	●	-	-	-	-	●
Poptella compressa	-	●	-	-	●	●
Prionobrama filigera	●	●	●	●	●	●
Pristobrycon sp.	-	-	●	-	-	-
Pygocentrus nattereri	-	-	●	-	-	●
Roeboides cf. *myersii*	-	-	●	-	-	-
Roeboides sp. 1	-	-	-	●	●	●
Roeboides sp. 2	-	●	-	●	-	●
Roeboides sp. 3	-	-	●	●	-	●
Serrasalminae sp.	-	-	-	-	-	●
Serrasalmus cf. *hollandi*	-	-	●	-	●	●
Serrasalmus marginatus	-	-	-	-	-	●
Serrasalmus rhombeus	●	-	●	-	-	●
Serrasalmus sp.	-	-	-	-	-	●
Stethaprion crenatum	-	●	-	-	-	●
Tetragonopterinae sp. 1	-	●	-	-	-	-
Tetragonopterinae sp. 2	●	-	-	●	-	-
Tetragonopterus argenteus	-	-	●	●	-	●
Triportheus angulatus	●	●	●	●	-	●
Triportheus sp.	●	-	●	-	-	-
Tyttocharax madeirae	●	●	-	●	-	-
Tyttocharax sp. nov.	●	-	-	-	-	-
Tyttocharax tambopatensis	●	-	●	-	-	-
Curimatidae						
Curimatella alburna	-	-	-	-	-	●
Curimatella dorsalis	-	-	-	-	-	●
Curimatella immaculata	-	●	-	-	-	●
Curimatella meyeri	-	●	●	-	-	●

TAXA	Upper Nareuda	Lower Nareuda	Upper Tahuamanu	Middle Tahuamanu	Lower Tahuamanu	Manuripi
Cyphocharax cf. *plumbeus*	-	-	-	-	-	●
Cyphocharax sp.	-	●	-	-	-	●
Cyphocharax spiluropsis	●	●	●	●	-	●
Potamorhina altamazonica	-	●	●	-	-	●
Potamorhina laitior	-	-	●	-	-	●
Psectrogaster curviventris	-	●	-	-	-	●
Psectrogaster rutiloides	-	-	-	-	-	●
Steindachnerina dobula	●	●	●	●	●	●
Steindachnerina guentheri	●	●	●	●	-	-
Steindachnerina leucisca	-	●	●	-	-	-
Steindachnerina sp.	●	●	-	●	-	-
Cynodontidae						
Cynodon gibbus	-	-	-	-	-	●
Hydrolycus pectoralis	●	-	●	-	-	-
Rhaphiodon vulpinus	●	-	●	-	-	-
Erythrinidae						
Hoplias malabaricus	●	●	●	●	-	●
Gasteropelecidae						
Carnegiella myersi	●	●	●	●	-	●
Carnegiella strigata	-	-	-	-	-	●
Gasteropelecus sternicla	●	●	●	●	-	●
Thoracocharax stellatus	-	-	●	●	●	●
Hemiodontidae						
Anodus elongatus	-	-	●	-	-	-
Lebiasinidae						
Nannostomus trifasciatus	-	-	-	-	-	●
Pyrrhulina australe	-	-	-	-	-	●
Pyrrhulina vittata	●	-	-	●	-	●
Prochilodontidae						
Prochilodus cf. *nigricans*	●	●	●	-	●	●
SILURIFORMES						
Ageneiosidae						
Ageneiosus cf. *caucanus*	-	-	-	-	-	●
Ageneiosus sp.	●	-	-	●	-	-
Tympanopleura sp.	-	-	-	-	●	●
Aspredinidae						
Bunocephalus coracoideus	-	-	-	-	-	●
Bunocephalus sp. 1	●	-	-	-	-	-
Bunocephalus sp. 2	-	●	-	-	-	●
Bunocephalus sp. 3	-	-	-	●	-	-

TAXA	Upper Nareuda	Lower Nareuda	Upper Tahuamanu	Middle Tahuamanu	Lower Tahuamanu	Manuripi
Dysichthys bifidus	-	●	-	-	-	●
Dysichthys cf. *aleuropsis*	-	●	●	●	-	-
Dysichthys cf. *amazonicus*	●	●	-	●	-	●
Dysichthys cf. *depressus*	●	●	-	-	-	-
Xiliphius cf. *melanopterus*	-	-	-	●	-	-
Auchenipteridae						
Auchenipterichthys thoracatus	-	-	-	-	-	●
Auchenipterus cf. *nuchalis*	-	-	●	-	-	-
Centromochlus cf. *heckelii*	-	-	-	●	-	-
Entomocorus benjamini	-	●	-	-	-	●
Tatia altae	●	●	●	●	-	-
Tatia aulopygia	-	-	-	-	-	●
Tatia cf. *perugiae*	-	●	-	-	-	-
Trachelyopterus cf. *galeatus*	-	-	-	-	-	●
Callichthyidae						
Brochis splendens	-	●	-	-	-	●
Callichthys callichthys	-	●	-	-	-	-
Corydoras acutus	●	●	●	●	-	●
Corydoras aeneus	●	-	-	-	-	●
Corydoras hastatus	-	-	-	-	-	●
Corydoras cf. *loretoensis*	●	●	●	●	●	●
Corydoras cf. *napoensis*	-	-	-	-	-	●
Corydoras sp.	-	-	●	-	-	●
Corydoras trilineatus	●	●	●	-	-	●
Dianema longibarbis	-	-	-	-	-	●
Megalechis thoractus	●	●	-	-	-	●
Cetopsidae						
Pseudocetopsis sp.	●	-	●	-	-	-
Doradidae						
Acanthodoras cataphractus	-	-	-	-	-	●
Agamyxis pectinifrons	-	-	-	-	-	●
Amblydoras cf. *hancockii*	-	-	-	-	-	●
Anadoras cf. *grypus*	-	-	-	-	-	●
Astrodoras asterifrons	-	-	-	-	-	●
Doras cf. *carinatus*	-	-	-	●	●	●
Doras eigenmanni	-	-	-	-	-	●
Hemidoras microstomus	-	-	-	-	-	●
Opsodoras cf. *humeralis*	-	-	-	-	-	●
Opsodoras cf. *stubelii*	-	-	-	●	-	●

TAXA	Upper Nareuda	Lower Nareuda	Upper Tahuamanu	Middle Tahuamanu	Lower Tahuamanu	Manuripi
Platydoras costatus	-	-	-	-	-	●
Pseudodoras niger	-	-	●	-	-	●
Trachydoras cf. *atripes*	-	-	-	●	-	-
Trachydoras paraguayensis	-	-	-	-	-	●
Loricariidae						
Ancistrus sp. 1	●	-	-	-	-	-
Ancistrus sp. 2	●	●	●	-	-	-
Ancistrus sp. 3	●	●	●	-	-	-
Ancistrus sp. 4	-	-	-	-	-	●
Aphanotorulus frankei	●	-	●	●	●	●
Aphanotorulus unicolor	-	-	-	●	-	-
Cochliodon cf. *cochliodon*	●	●	●	●	-	●
Crossoloricaria sp.	-	-	●	●	●	●
Farlowella cf. *oxyrryncha*	-	-	●	●	-	●
Farlowella sp. 1	●	●	-	●	-	●
Farlowella sp. 2	●	-	●	●	-	●
Glyptoperichthys lituratus	-	-	-	-	-	●
Hemiodontichthys acipenserinus	-	●	●	-	-	●
Hypoptopoma joberti	-	●	●	-	-	●
Hypoptopoma sp.	●	●	●	-	-	●
Hypostomus sp. 1	●	●	●	-	-	●
Hypostomus sp. 2	●	●	●	●	-	●
Hypostomus sp. 3	-	-	-	●	-	●
Hypostomus sp. 4	-	-	-	-	-	●
Lamontichthys filamentosus	-	-	●	●	-	-
Liposarcus disjunctivus	-	-	●	-	-	●
Loricaria sp.	●	●	●	●	●	●
Loricariidae sp.	●	-	-	-	-	-
Loricariichthys sp.	-	●	●	●	-	●
Otocinclus mariae	●	●	●	●	-	●
Panaque sp.	-	-	-	●	-	-
Parotocinclus sp.	-	●	●	-	-	●
Peckoltia arenaria	-	●	-	●	-	-
Planiloricaria cryptodon	-	-	-	●	-	-
Pseudohemiodon cf. *lamina*	-	-	-	●	-	●
Pseudohemiodon sp. 1	-	-	-	-	-	●
Pseudohemiodon sp. 2	-	-	-	●	-	-
Pseudohemiodon sp. 3	●	-	-	-	-	-
Rineloricaria lanceolata	●	●	●	●	-	●

TAXA	Upper Nareuda	Lower Nareuda	Upper Tahuamanu	Middle Tahuamanu	Lower Tahuamanu	Manuripi
Rineloricaria sp.	●	●	●	●	-	●
Scoloplax cf. *dicra*	-	-	-	-	-	●
Sturisoma nigrirostrum	●	●	●	●	-	●
Pimelodidae						
Brachyglanis? sp.	-	●	-	-	-	-
Brachyrhamdia marthae	-	●	●	-	-	●
Cetopsorhamdia phantasia	-	-	-	●	-	-
Cheirocerus eques	-	-	●	●	●	●
Duopalatinus cf. *malarmo*	-	-	●	-	-	-
Hemisorubim platyrhynchos	●	-	-	-	-	●
Heptapterus longior	●	-	-	-	-	-
Heptapterus sp.	●	-	-	-	-	-
Imparfinis sp.	●	-	-	-	-	-
Imparfinis stictonotus	●	●	●	●	●	●
Leiarius marmoratus	-	-	●	●	-	●
Megalonema sp.	●	-	-	●	-	●
Megalonema sp. nov.	-	-	●	●	●	●
Microglanis sp.	●	●	●	-	-	-
Pimelodella cf. *boliviana*	-	-	-	-	-	●
Pimelodella cristata	●	●	-	-	●	●
Pimelodella gracilis	●	●	●	●	●	●
Pimelodella hasemani	-	-	●	●	-	●
Pimelodella cf. *itapicuruensis*	●	●	-	●	-	●
Pimelodella cf. *serrata*	-	●	●	●	●	●
Pimelodidae sp.	-	-	-	-	-	●
Pimelodus "*altipinnis*"	-	-	-	●	-	●
Pimelodus altissimus (sp. nov.)	-	-	-	●	-	-
Pimelodus armatus	●	-	-	-	-	-
Pimelodus cf. *blochii*	-	-	●	●	●	●
Pimelodus cf. *pantherinus*	-	●	●	●	-	●
Pimelodus sp. 1	-	-	-	-	-	●
Pimelodus sp. 2	-	-	-	●	-	-
Pimelodus sp. 3	-	-	-	-	-	●
Pimelodus sp. 4	-	-	-	●	-	-
Pseudoplatystoma fasciatum	-	-	-	-	-	●
Rhamdia sp.	-	●	-	-	-	●
Sorubim lima	-	●	●	-	-	●
Trichomycteridae						
Acanthopoma cf. *bondi*	-	-	-	●	-	-
Homodiaetus sp.	●	-	●	-	●	●

TAXA	Upper Nareuda	Lower Nareuda	Upper Tahuamanu	Middle Tahuamanu	Lower Tahuamanu	Manuripi
Ochmacanthus cf. *alternus*	●	●	●	●	-	●
Plectrochilus sp.	-	-	-	●	-	-
Pseudostegophilus nemurus	-	-	●	-	●	●
Tridentopsis pearsoni	-	-	-	-	-	●
Vandellia cirrhosa	●	-	●	●	●	●
GYMNOTIFORMES						
Apteronotidae						
Adontosternarchus clarkae	-	-	-	-	-	●
Apteronotus albifrons	-	●	●	-	-	●
Apteronotus bonapartii	-	-	-	●	-	●
Electrophoridae						
Electrophorus electricus	-	-	-	-	-	●*
Gymnotidae						
Gymnotus cf. *anguillaris*	●	-	-	-	-	-
Gymnotus carapo	-	-	●	-	-	●
Gymnotus cf. *coatesi*	●	●	-	●	-	-
Hypopomidae						
Brachyhypopomus brevirostris	-	-	-	-	-	●
Brachyhypopomus pinnicaudatus	-	-	-	-	-	●
Brachyhypopomus sp.	-	●	-	-	-	●
Hypopygus lepturus	-	-	-	-	-	●
Sternopygidae						
Distocyclus conirostris	-	-	-	-	-	●
Eigenmannia humboldtii	-	-	-	-	-	●
Eigenmannia macrops	●	-	-	●	-	●
Eigenmannia cf. *trilineata*	-	●	-	-	-	●
Eigenmannia virescens	●	●	●	●	●	●
Rhabdolichops caviceps	-	-	-	-	-	●
Sternopygus macrurus	-	-	●	-	-	●
CYPRINODONTIFORMES						
Rivulidae						
Rivulus sp.	-	-	-	-	-	●
BELONIFORMES						
Belonidae						
Potamorrhaphis sp.	●	-	-	●	-	-
SYNBRANCHIFORMES						
Synbranchidae						
Synbranchus marmoratus	●	-	-	-	-	●

TAXA	Upper Nareuda	Lower Nareuda	Upper Tahuamanu	Middle Tahuamanu	Lower Tahuamanu	Manuripi
PERCIFORMES						
Cichlidae						
Aequidens sp. 1	●	●	●	-	-	●
Aequidens sp. 2	●	●	●	●	-	●
Aequidens sp. 3	-	-	●	-	-	●
Aequidens cf. *paraguayensis*	●	●	●	●	-	-
Aequidens cf. *tetramerus*	-	●	-	-	-	-
Apistogramma linkei	●	●	●	-	-	●
Apistogramma sp. 1	●	-	●	-	-	●
Apistogramma sp. 2	●	●	-	●	●	●
Apistogramma sp. 3	-	●	-	-	-	-
Apistogramma sp. 4	-	-	-	-	-	●
Astronotus crassipinnis	-	-	-	-	-	●
Chaetobranchiopsis orbicularis	-	●	●	-	-	●
Cichla cf. *monoculus*	-	-	-	-	-	●
Cichlasoma severum	-	-	-	-	-	●
Cichlidae sp.	●	-	●	-	-	-
Crenicara cf. *unctulata*	-	-	-	-	-	●
Crenicichla cf. *heckell*	●	-	●	●	●	●
Crenicichla sp. 1	-	-	-	●	-	-
Crenicichla sp. 2	●	●	-	-	●	●
Mesonauta festivus	-	-	-	●	-	●
Mesonauta cf. *insignis*	-	-	-	-	-	●
Microgeophagus altispinosa	-	-	-	-	-	●
Satanoperca cf. *acuticeps*	-	-	-	-	-	●
Satanoperca sp.	-	-	●	●	-	●
Sciaenidae						
Pachyurus sp.	-	-	-	-	●	●
Plagioscion squamosissimus	-	-	●	-	-	-

APPENDIX 7 Comparative list of fishes reported from the Bolivian Amazon.

Philip W. Willink, Jaime Sarmiento, and Barry Chernoff

See text at end of table for explanation of localities and sources of information.

TAXA	Bolivia AquaRAP	Noel Kempff Mercado	Bolivian Amazon	Madre de Dios	Itenez (or Guapore)
RAJIFORMES					
Potamotrygonidae					
Potamotrygon cf. *hystrix*	-	-	●	-	-
Potamotrygon motoro	●	-	-	-	-
Potamotrygon cf. *motoro*	-	-	●	●	●
Potamotrygon sp.	-	-	●	-	●
Potamotrygonidae sp.	-	-	●	-	●
LEPIDOSIRENIFORMES					
Lepidosirenidae					
Lepidosiren paradoxa	-	●	●	-	-
CLUPEIFORMES					
Clupeidae					
Pellona castelnaeana	-	-	●	●	●
Pellona flavipinnis	-	-	●	●	●
Engraulidae					
Anchoviella cf. *carrikeri*	●	-	-	-	-
Engraulidae sp. 1	-	-	●	-	●
Engraulidae sp. 2	-	-	●	●	-
CHARACIFORMES					
Anostomidae					
Abramites hypselonotus	●	-	●	-	-
Anostomus cf. *gracilis*	-	-	●	-	●
Anostomus cf. *plicatus*	-	-	●	-	●
Anostomus proximus	-	-	●	-	●
Anostomus taeniatus	-	-	●	-	●
Laemolyta sp. **	●	-	-	-	-
Leporinus cf. *cylindriformis*	-	●	-	-	-
Leporinus fasciatus	-	-	●	-	●
Leporinus cf. *fasciatus*	●	-	-	-	-
Leporinus friderici	●	-	●	-	●
Leporinus cf. *friderici*	-	●	-	-	-
Leporinus cf. *nattereri* **	●	-	-	-	-
Leporinus pearsoni	-	-	●	●	-
Leporinus striatus	-	-	●	-	-

TAXA	Bolivia AquaRAP	Noel Kempff Mercado	Bolivian Amazon	Madre de Dios	Itenez (or Guapore)
Leporinus trifasciatus	-	-	●	●	-
Leporinus sp. nov. (*amazonensis*)	-	-	●	-	-
Rhytiodus argenteofuscus	-	-	●	●	●
Rhytiodus lauzannei	-	-	●	-	-
Rhytiodus microlepis	-	-	●	-	●
Schizodon fasciatum	●	●	●	●	●
Characidae					
Acestrorhynchus altus	-	-	●	-	●
Acestrorhynchus falcatus	-	-	●	●	-
Acestrorhynchus falcirostris	-	-	●	-	●
Acestrorhynchus heterolepis	-	-	●	-	●
Acestrorhynchus microlepis	-	-	●	-	●
Acestrorhynchus cf. *minimus*	-	-	●	-	●
Acestrorhynchus sp.	-	●	-	-	-
Aphyodite cf. *grammica*	-	-	●	●	-
Aphyocharax alburnus	●	●	●	●	-
Aphyocharax dentatus	●	-	●	-	-
Aphyocharax paraguayensis	-	-	●	-	-
Aphyocharax pusillus **	●	-	-	-	-
Aphyocharax rathbuni	-	●	-	-	-
Aphyocheirodon sp. nov.	-	-	●		-
Astyanacinus cf. *moori*	-	-	●	-	-
Astyanacinus multidens	-	-	●	-	-
Astyanax abramis	-	-	●	-	-
Astyanax cf. *abramis*	●	-	-	-	-
Astyanax bimaculatus	-	-	●	-	●
Astyanax cf. *daguae*	-	-	●	-	●
Astyanax fasciatus	-	-	●	-	-
Astyanax lineatus	-	-	●	-	-
Astyanax cf. *mucronatus*	-	-	●	-	-
Astyanax sp.	●	-	-	-	-
Astyanax sp. 1	-	●	-	-	-
Astyanax sp. 2	-	●	-	-	-
Brachychalcinus copei	●	-	●	-	-
Brachychalcinus orbicularis	-	●	●	●	●
Brycon cephalus	-	-	●	●	●
Brycon erythropterum	-	-	●	-	-
Bryconacidnus ellisi	-	-	●	-	-
Bryconamericus bolivianus	-	-	●	-	-
Bryconamericus cf. *caucanus* **	●	-	-		-
Bryconamericus cf. *pachacuti* **	●	-	-	-	-
Bryconamericus cf. *peruanus*	●	-	-	-	-
Bryconamericus cf. *peruanus*	-	-	●	-	-

TAXA	Bolivia AquaRAP	Noel Kempff Mercado	Bolivian Amazon	Madre de Dios	Itenez (or Guapore)
Bryconamericus sp.	●	-	-	-	-
Bryconops cf. *alburnoides*	-	-	●	-	●
Bryconops cf. *caudomaculatus*	-	●	-	-	-
Bryconops melanurus	-	●	●	-	●
Bryconops sp.	-	●	-	-	-
Bryconops sp.	-	-	●	●	●
Catoprion mento	-	-	●	-	●
Chalceus erythrurus	-	-	●	●	-
Chalceus macrolepidotus	-	-	●	-	●
Characidium bolivianum	-	-	●	-	-
Characidium cf. *fasciatum*	-	●	-	-	-
Characidium sp. 1	●	-	-	-	-
Characidium sp. 2	●	-	-	-	-
Charax gibbosus	●	●	●	-	●
Cheirodon fugitiva	●	-	-	-	-
Cheirodon cf. *fugitiva*	-	-	●	●	-
Cheirodon sp. 1	●	-	-	-	-
Cheirodon sp. 2	●	-	-	-	-
Cheirodon spp.	-	●	-	-	-
Cheirodon sp.	-	-	●	-	-
Cheirodontinae sp.	●	-	-	-	-
Cheirodontinae sp.	-	●	-	-	-
Cheirodontinae sp. 1	-	-	●	-	-
Cheirodontinae sp. 2	-	-	●	-	-
Cheirodontinae sp. (gr. *Aphyodite*)	-	●	-	-	-
Chrysobrycon sp. 1 **	●	-	-	-	-
Chrysobrycon sp. 2 **	●	-	-	-	-
Clupeacharax anchoveoides	●	●	-	-	-
Colossoma macropomum	-	-	●	●	●
Creagrutus beni	-	-	●	-	-
Creagrutus sp. 1	●	-	-	-	-
Creagrutus sp. 2	●	-	-	-	-
Creagrutus sp. 3	●	-	-	-	-
Ctenobrycon spilurus	●	●	●	●	●
Cynopotamus amazonus	-	-	●	-	-
Cynopotamus gouldingi **	●	-	-	-	-
Engraulisoma taeniatum **	●	-	-	-	-
Eucynopotamus biserialis **	●	-	-	-	-
Eucynopotamus sp. 1	-	-	●	-	-
Eucynopotamus sp. 2	-	-	●	-	●
Galeocharax gulo	●	-	●	●	-
Gephyrocharax chaparae	-	-	●	-	-
Gephyrocharax major ++	-	-	-	-	-
Gephyrocharax sp.	●	-	-	-	-

TAXA	Bolivia AquaRAP	Noel Kempff Mercado	Bolivian Amazon	Madre de Dios	Itenez (or Guapore)
Gnathocharax steindachneri	-	●	-	-	-
Gymnocorymbus ternetzi	-	-	●	-	●
Gymnocorymbus thayeri	-	-	●	-	-
Hemibrycon sp.	-	-	●	-	-
Hemigrammus cf. *bellottii*	-	●	-	-	-
Hemigrammus lunatus	●	●	●	●	●
Hemigrammus cf. *marginatus*	-	-	●	-	-
Hemigrammus cf. *megaceps* **	●	-	-	-	-
Hemigrammus ocellifer	●	●	-	-	-
Hemigrammus cf. *pretoensis* **	●	-	-	-	-
Hemigrammus sp.	-	-	●	-	-
Hemigrammus? sp.	●	-	-	-	-
Hemigrammus cf. *tridens*	-	●	-	-	-
Hemigrammus unilineatus	-	-	●	-	-
Hemigrammus cf. *unilineatus*	●	●	-	-	-
Holobrycon pesu	-	-	●	-	●
Hyphessobrycon agulha **	●	-	-	-	-
Hyphessobrycon cf. *anisitsi* **	●	-	-	-	-
Hyphessobrycon bentosi	-	-	●	-	-
Hyphessobrycon cf. *bentosi*	-	●	-	-	-
Hyphessobrycon eques	-	●	●	-	-
Hyphessobrycon cf. *gracilior* **	●	-	-	-	-
Hyphessobrycon cf. *herbertaxelrodi*	-	●	-	-	-
Hyphessobrycon cf. *heterorhabdus*	-	●	-	-	-
Hyphessobrycon megalopterus	-	●	-	-	-
Hyphessobrycon cf. *minimus*	-	●	-	-	-
Hyphessobrycon cf. *scholzei*	-	●	-	-	-
Hyphessobrycon serpae	-	-	●	●	-
Hyphessobrycon? sp.	●	-	-	-	-
Hyphessobrycon sp.	-	●	-	-	-
Hyphessobrycon sp.	-	-	●	●	-
Hyphessobrycon cf. *tucunai*	●	●	-	-	-
Iguanodectes spilurus	●	●	●	-	●
Jobertina lateralis	-	●	-	-	-
Knodus breviceps	-	-	●	-	-
Knodus cf. *caquetae* **	●	-	-	-	-
Knodus cf. *gamma* **	●	-	-	-	-
Knodus cf. *heterestes* **	●	-	-	-	-
Knodus cf. *moenkhausii*	-	-	●	-	-
Knodus sp.	●	-	-	-	-
Knodus sp.	-	●	-	-	-
Knodus sp. 1	-	-	●	-	-
Knodus sp. 2	-	-	●	-	●
Knodus cf. *victoriae* **	●	-	-	-	-

TAXA	Bolivia AquaRAP	Noel Kempff Mercado	Bolivian Amazon	Madre de Dios	Itenez (or Guapore)
Markiana nigripinnis	-	-	●	-	●
Metynnis argenteus	-	-	●	-	-
Metynnis hypsauchen	-	-	●	-	●
Metynnis cf. *hypsauchen*	-	-	●	-	●
Metynnis gr. *lippincotianus*	-	-	●	-	-
Metynnis luna **	●	-	-	-	-
Metynnis gr. *maculatus* 1	-	-	●	-	●
Metynnis gr. *maculatus* 2	-	-	●	-	●
Microschemobrycon geisleri **	●	-	-	-	-
Microschemobrycon hasemani	-	-	●	●	-
Moenkhausia cf. *chrysargyrea* **	●	-	-	-	-
Moenkhausia colletti	●	●	-	-	-
Moenkhausia cf. *colletti*	-	-	●	-	●
Moenkhausia cf. *comma* **	●	-	-	-	-
Moenkhausia cf. *cotinho*	-	-	●	-	●
Moenkhausia dichroura	●	●	●	●	●
Moenkhausia grandisquamis	-	-	●	-	●
Moenkhausia jamesi	-	-	●	●	●
Moenkhausia cf. *jamesi*	●	-	-	-	-
Moenkhausia cf. *lepidura*	●	●	-	-	-
Moenkhausia cf. *lepidura*	-	-	●	-	●
Moenkhausia cf. *megalops* **	●	-	-	-	-
Moenkhausia oligolepis	-	●	●	-	●
Moenkhausia sanctaefilomenae	●	●	●	-	-
Moenkhausia sp. 1	●	-	-	-	-
Moenkhausia sp. 2	●	-	-	-	-
Moenkhausia sp. 3	●	-	-	-	-
Moenkhausia sp. 4	●	-	-	-	-
Moenkhausia sp. 5	●	-	-	-	-
Moenkhausia sp. 6	●	-	-	-	-
Moenkhausia sp. 7	●	-	-	-	-
Moenkhausia sp. 8	●	-	-	-	-
Moenkhausia sp.	-	●	-	-	-
Myleus sp.	●	-	-	-	-
Myleus cf. *rubripinnis*	-	-	●	-	-
Myleus tiete	-	-	●	-	●
Mylossoma aureum	-	-	●	-	-
Mylossoma duriventris	●	-	●	●	●
Odontostilbe hasemani	●	-	●	●	-
Odontostilbe piaba	●	-	●	-	-
Odontostilbe paraguayensis **	●	-	-	-	-
Odontostilbe sp. 1	●	-	-	-	-
Odontostilbe sp. 2	●	-	-	-	-
Odontostilbe sp.	-	-	●	-	-

TAXA	Bolivia AquaRAP	Noel Kempff Mercado	Bolivian Amazon	Madre de Dios	Itenez (or Guapore)
Odontostilbe stenodon	-	-	●	●	-
Parecbasis cyclolepis	-	-	●	●	●
Paragoniates alburnus	●	-	●	-	-
Phenacogaster cf. *microstictus* **	●	-	-	-	-
Phenacogaster cf. *pectinatus* **	●	-	-	-	-
Phenacogaster sp. 1	●	-	-	-	-
Phenacogaster sp. 2	●	-	-	-	-
Phenacogaster sp. 3	●	-	-	-	-
Phenacogaster? sp.	●	-	-	-	-
Phenacogaster sp.	-	●	-	-	-
Phenacogaster sp.	-	-	●	-	●
Phenacogaster sp. nov.	-	-	●	-	-
Piabucus melanostomus	●	-	●	-	●
Piaractus brachypomus	-	-	●	●	-
Poptella compressa **	●	-	-	-	-
Prionobrama filigera	●	-	●	●	-
Pristobrycon sp. **	●	-	-	-	-
Prodontocharax melanotus	-	-	●	-	-
Pseudocheirodon sp.	-	●	-	-	-
Pygocentrus nattereri	●	-	●	●	●
Roeboides cf. *descalvadensis*	-	-	●	-	-
Roeboides myersii	-	-	●	●	●
Roeboides cf. *myersii*	●	-	-	-	-
Roeboides sp. 1	●	-	-	-	-
Roeboides sp. 2	●	-	-	-	-
Roeboides sp. 3	●	-	-	-	-
Roeboides sp.	-	-	●	-	●
Roestes molossus	-	-	●	-	●
Salminus affinis	-	-	●	-	-
Salminus brasiliensis	-	-	●	-	-
Serrasalminae sp.	●	-	-	-	-
Serrasalmus compressus	-	-	●	●	●
Serrasalmus eigenmanni	-	-	●	●	●
Serrasalmus elongatus	-	-	●	-	●
Serrasalmus hollandi	-	-	●	-	●
Serrasalmus cf. *hollandi*	●	-	-	-	-
Serrasalmus marginatus **	●	-	-	-	-
Serrasalmus rhombeus	●	-	●	●	●
Serrasalmus sp.	●	-	-	-	-
Serrasalmus sp.	-	●	-	-	-
Serrasalmus spilopleura	-	-	●	●	●
Stethaprion crenatum	●	●	●	-	●
Tetragonopterinae sp. 1	●	-	-	-	-
Tetragonopterinae sp. 2	●	-	-	-	-

Comparative list of fishes reported from the Bolivian Amazon

TAXA	Bolivia AquaRAP	Noel Kempff Mercado	Bolivian Amazon	Madre de Dios	Itenez (or Guapore)
Tetragonopterus argenteus	●	-	●	●	●
Tetragonopterus cf. *chalceus*	-	-	●	-	●
Thayeria boehlkei	-	●	●	-	●
Triportheus albus	-	-	●	●	●
Triportheus angulatus	●	●	●	●	●
Triportheus culter	-	-	●	-	●
Triportheus sp.	●	-	-	-	-
Tyttocharax madeirae	●	●	●	-	-
Tyttocharax sp. nov.	●	-	-	-	-
Tyttocharax tambopatensis **	●	-	-	-	-
Xenurobrycon polyancistrus	-	-	●	-	-
Curimatidae					
Chilodus punctatus	-	●	●	-	●
Curimata roseni	-	-	●	-	●
Curimata vittata	-	-	●	-	●
Curimatella alburna	●	-	●	-	●
Curimatella dorsalis	●	-	●	-	●
Curimatella immaculata	●	-	●	-	●
Curimatella meyeri	●	-	●	●	●
Curimatopsis macrolepis	-	●	-	-	-
Cyphocharax notatus	-	●	-	-	-
Cyphocharax cf. *plumbeus*	●	-	-	-	-
Cyphocharax cf. *plumbeus*	-	-	●	-	●
Cyphocharax sp.	●	-	-	-	-
Cyphocharax sp. nov.	-	-	●	-	●
Cyphocharax spiluropsis	●	-	-	-	-
Cyphocharax cf. *spiluropsis*	-	●	-	-	-
Cyphocharax cf. *spilurus*	-	●	-	-	-
Cyphocharax cf. *spilurus*	-	-	●	●	●
Eigenmannina melanopogon	-	-	●	●	●
Potamorhina altamazonica	●	-	●	●	●
Potamorhina laitior	●	-	●	●	●
Psectrogaster curviventris	●	-	●	-	●
Psectrogaster essequibensis	-	-	●	-	●
Psectrogaster rutiloides	●	●	●	-	●
Steindachnerina bimaculata	-	-	●	-	-
Steindachnerina binotata	-	-	●	-	-
Steindachnerina dobula	●	-	●	●	-
Steindachnerina hypostoma	-	-	●	●	-
Steindachnerina guentheri **	●	-	-	-	-
Steindachnerina leucisca **	●	-	-	-	-
Steindachnerina sp.	●	-	-	-	-
Steindachnerina sp.	-	●	-	-	-

TAXA	Bolivia AquaRAP	Noel Kempff Mercado	Bolivian Amazon	Madre de Dios	Itenez (or Guapore)
Cynodontidae					
Cynodon gibbus	●	-	●	●	●
Hydrolycus cf. *armatus*	-	-	●	●	-
Hydrolycus pectoralis **	●	-	-	-	-
Hydrolycus scomberoides	-	-	●	●	●
Rhaphiodon vulpinus	●	-	●	●	-
Erythrinidae					
Erythrinus erythrinus	-	-	●	-	-
Erythrinus sp.	-	●	-	-	-
Hoplias malabaricus	●	●	●	●	●
Hoplerythrinus unitaeniatus	-	●	●	-	-
Gasteropelecidae					
Carnegiella marthae	-	●	-	-	-
Carnegiella myersi	●	-	●	-	-
Carnegiella schereri	-	-	●	-	●
Carnegiella strigata	●	●	-	-	-
Gasteropelecus sternicla	●	●	●	-	-
Thoracocharax securis	-	-	●	-	-
Thoracocharax stellatus	●	-	●	-	-
Hemiodontidae					
Anodus elongatus **	●	-	-	-	-
Hemiodopsis cf. *microlepis*	-	-	●	-	●
Hemiodopsis semitaeniatus	-	-	●	-	●
Hemiodus unimaculatus	-	-	●	●	●
Parodon cf. *carrikeri*	-	-	●	-	-
Lebiasinidae					
Nannostomus sp.	-	●	-	-	-
Nannostomus harrisoni	-	●	-	-	-
Nannostomus trifasciatus	●	●	-	-	-
Nannostomus unifasciatus	-	-	●	-	●
Pyrrhulina australe	●	●	●	-	●
Pyrrhulina brevis	-	●	-	-	-
Pyrrhulina vittata	●	-	●	-	●
Prochilodontidae					
Prochilodus nigricans	-	-	●	●	●
Prochilodus cf. *nigricans*	●	-	-	-	-
Prochilodus sp. 1	-	-	●	-	●
Prochilodus sp. 2	-	-	●	-	-
SILURIFORMES					
Ageneiosidae					
Ageneiosus brevifilis	-	-	●	●	-
Ageneiosus cf. *caucanus* **	●	-	-	-	-
Ageneiosus dentatus	-	-	●	●	-

APPENDIX 7

Comparative list of fishes reported from the Bolivian Amazon

TAXA	Bolivia AquaRAP	Noel Kempff Mercado	Bolivian Amazon	Madre de Dios	Itenez (or Guapore)
Ageneiosus madeirensis	-	-	●	-	●
Ageneiosus sp.	●	-	-	-	-
Ageneiosus sp.	-	-	●	-	-
Ageneiosus ucayalensis	-	-	●	-	-
Tympanopleura sp.	●	-	-	-	-
Tympanopleura sp.	-	-	●	-	●
Aspredinidae					
Amaralia sp.	-	-	●	-	-
Bunocephalus coracoideus **	●	-	-	-	-
Bunocephalus sp. 1	●	-	-	-	-
Bunocephalus sp. 2	●	-	-	-	-
Bunocephalus sp. 3	●	-	-	-	-
Bunocephalus sp.	-	●	-	-	-
Bunocephalus sp. 1	-	-	●	-	●
Bunocephalus sp. 2	-	-	●	-	-
Bunocephalus sp. 3	-	-	●	-	-
Dysichthys bifidus **	●	-	-	-	-
Dysichthys cf. *aleuropsis* **	●	-	-	-	-
Dysichthys cf. *amazonicus* **	●	-	-	-	-
Dysichthys cf. *depressus* **	●	-	-	-	-
Xiliphius cf. *melanopterus* **	●	-	-	-	-
Astroblepidae					
Astroblepus longiceps	-	-	●	-	-
Astroblepus sp.	-	-	●	-	-
Auchenipteridae					
Auchenipterichthys thoracatus	●	-	●	-	●
Auchenipterus nigripinnis	-	-	●	●	●
Auchenipterus nuchalis	-	-	●	●	●
Auchenipterus cf. *nuchalis*	●	-	-	-	-
Centromochlus cf. *heckelii* **	●	-	-	-	-
Centromochlus sp. 1	-	-	●	-	-
Centromochlus sp. 2	-	-	●	-	-
Entomocorus benjamini	●	-	●	-	●
Epapterus dispilurus	-	-	●	-	●
Pseudotatia? sp.	-	●	-	-	-
Tatia altae **	●	-	-	-	-
Tatia aulopygia	●	●	●	-	-
Tatia cf. *intermedia*	-	●	-	-	-
Tatia cf. *perugiae* **	●	-	-	-	-
Tatia sp.	-	●	-	-	-
Trachelyopterus coriaceus	-	-	●	-	-
Trachelyopterus cf. *galeatus*	●	-	-	-	-
Trachelyopterus cf. *galeatus*	-	-	●	-	-
Trachelyopterus maculosus	-	-	●	-	-

TAXA	Bolivia AquaRAP	Noel Kempff Mercado	Bolivian Amazon	Madre de Dios	Itenez (or Guapore)
Trachelyopterus striatulus	-	-	●	-	-
Callichthyidae					
Brochis britskii	-	-	●	-	-
Brochis multiradiatus	-	-	●	-	-
Brochis splendens	●	-	●	-	-
Callichthys callichthys	●	●	●	-	-
Corydoras acutus	●	-	●	-	-
Corydoras aeneus	●	●	●	-	-
Corydoras armatus	-	-	●	-	-
Corydoras bolivianus	-	-	●	-	-
Corydoras geryi	-	-	●	-	-
Corydoras hastatus	●	●	●	-	-
Corydoras cf. *latus*	-	-	●	-	-
Corydoras cf. *loretoensis* **	●	-	-	-	-
Corydoras cf. *napoensis* **	●	-	-	-	-
Corydoras punctatus	-	-	●	-	-
Corydoras sp.	●	-	-	-	-
Corydoras sp.	-	●	-	-	-
Corydoras sp. 1	-	-	●	-	●
Corydoras sp. 2	-	-	●	-	-
Corydoras trilineatus **	●	-	-		-
Dianema longibarbis	●	-	●	-	●
Hoplosternum littoralis	-	●	●	●	-
Megalechis thoractus	●	-	●	-	-
Cetopsidae					
Cetopsis sp.	-	-	●	-	-
Hemicetopsis candiru	-	-	●	-	-
Pseudocetopsis plumbeus	-	-	●	-	-
Pseudocetopsis sp.	●	-	-	-	-
Pseudocetopsis sp.	-	-	●	-	-
Doradidae					
Acanthodoras cataphractus **	●	-	-	-	-
Acanthodoras spinosissimus	-	●	-	-	-
Agamyxis flavopictus	-	-	●	-	-
Agamyxis pectinifrons **	●	-	-	-	-
Amblydoras hancockii	-	●	-	-	-
Amblydoras cf. *hancockii*	●	-	-	-	-
Anadoras cf. *grypus* **	●	-	-	-	-
Anadoras weddellii	-	-	●	-	-
Astrodoras asterifrons	●	-	●		-
Doras cf. *carinatus* **	●	-	-	-	-
Doras eigenmanni	●	-	●	-	-
Doras fimbriatus	-	-	●	-	-

TAXA	Bolivia AquaRAP	Noel Kempff Mercado	Bolivian Amazon	Madre de Dios	Itenez (or Guapore)
Doras punctatus	-	-	●	-	●
Doras sp.	-	-	●	-	-
Hemidoras microstomus **	●	-	-	-	-
Megalodoras irwini	-	-	●	-	-
Opsodoras humeralis	-	-	●	-	-
Opsodoras cf. *humeralis*	●	-	-	-	-
Opsodoras stubelii	-	-	●	-	-
Opsodoras cf. *stubelii*	●	-	-	-	-
Opsodoras sp. 1	-	-	●	-	-
Opsodoras sp. 2	-	-	●	-	●
Platydoras costatus	●	-	●	-	●
Pseudodoras niger	●	-	●	●	●
Pterodoras granulosus	-	-	●	●	-
Trachydoras atripes	-	-	●	-	●
Trachydoras cf. *atripes*	●	-	-	-	-
Trachydoras paraguayensis	●	-	-	-	-
Trachydoras cf. *paraguayensis*	-	-	●	-	-
Helogenidae					
Helogenes marmoratus	-	●	-	-	-
Hypophthalmidae					
Hypophthalmus edentatus	-	-	●	●	-
Hypophthalmus marginatus	-	-	●	-	-
Loricariidae					
Ancistrus cf. *bolivianus*	-	-	●	-	-
Ancistrus cf. *megalostomus*	-	-	●	-	-
Ancistrus sp. 1	●	-	-	-	-
Ancistrus sp. 2	●	-	-	-	-
Ancistrus sp. 3	●	-	-	-	-
Ancistrus sp. 4	●	-	-	-	-
Ancistrus sp.	-	●	-	-	-
Ancistrus sp.	-	-	●	-	-
Ancistrus cf. *temminckii*	-	-	●	●	●
Aphanotorulus frankei	●	-	●	-	-
Aphanotorulus cf. *popoi*	-	-	●	●	●
Aphanotorulus unicolor **	●	-	-	-	-
Cochliodon cf. *cochliodon* **	●	-	-	-	-
Cochliodon sp. 1	-	-	●	-	-
Cochliodon sp. 2	-	-	●	-	-
Crossoloricaria sp. **	●	-	-	-	-
Farlowella nattereri	-	-	●	-	-
Farlowella cf. *oxyrryncha* **	●	-	-	-	-
Farlowella sp. 1	●	-	-	-	-
Farlowella sp. 2	●	-	-	-	-
Farlowella sp.	-	●	-	-	-

TAXA	Bolivia AquaRAP	Noel Kempff Mercado	Bolivian Amazon	Madre de Dios	Itenez (or Guapore)
Farlowella sp. 1	-	-	●	-	-
Farlowella sp. 2	-	-	●	-	-
Glyptoperichthys lituratus	●	-	●	●	●
Glyptoperichthys punctatus	-	-	●	-	-
Hemiodontichthys acipenserinus	●	-	●	-	●
Hypoptopoma joberti	●	●	●	●	●
Hypoptopoma sp.	●	-	-	-	-
Hypoptopoma thoracatum	-	●	●	-	●
Hypoptopomatinae sp.	-	●	-	-	-
Hypoptopomatinae sp. nov. ?	-	-	●	-	-
Hypostomus bolivianus	-	-	●	-	-
Hypostomus cf. *chaparae*	-	-	●	-	-
Hypostomus emarginatus	-	-	●	-	●
Hypostomus sp. 1	●	-	-	-	-
Hypostomus sp. 2	●	-	-	-	-
Hypostomus sp. 3	●	-	-	-	-
Hypostomus sp. 4	●	-	-	-	-
Hypostomus sp.	-	●	-	-	-
Hypostomus sp. 2	-	●	-	-	-
Hypostomus sp. 3	-	●	-	-	-
Hypostomus sp. 1	-	-	●	-	-
Hypostomus sp. 2	-	-	●	-	-
Hypostomus sp. 3	-	-	●	-	-
Hypostomus sp. 4	-	-	●	-	●
Hypostomus sp. 5	-	-	●	-	-
Lamontichthys filamentosus	●	-	-	-	-
Lamontichthys cf. *filamentosus*	-	-	●	-	-
Liposarcus disjunctivus	●	-	-	-	-
Liposarcus disjunctivus (sp. nov.)	-	-	●	-	-
Loricaria cf. *simillima*	-	-	●	-	-
Loricaria sp.	●	-	-	-	-
Loricaria sp.	-	●	-	-	-
Loricariidae sp.	●	-	-	-	-
Loricariinae sp.	-	●	-	-	-
Loricariichthys cf. *maculatus*	-	-	●	●	●
Loricariichthys sp.	●	-	-	-	-
Loricariichthys sp.	-	-	●	●	●
Otocinclus mariae	●	-	●	-	-
Otocinclus cf. *mariae*	-	●	-	-	-
Panaque sp.	●	-		-	-
Panaque sp. n 1	-	-	●	-	-
Panaque sp. n 2	-	-	●	-	-
Parotocinclus sp. **	●	-	-	-	-
Peckoltia arenaria	●	-	-	-	-

TAXA	Bolivia AquaRAP	Noel Kempff Mercado	Bolivian Amazon	Madre de Dios	Itenez (or Guapore)
Peckoltia cf. *arenaria*	-	-	●	-	-
Planiloricaria cryptodon	●	-	●	-	-
Pseudohemiodon cf. *lamina* **	●	-	-	-	-
Pseudohemiodon sp. 1	●	-	-	-	-
Pseudohemiodon sp. 2	●	-	-	-	-
Pseudohemiodon sp. 3	●	-	-	-	-
Pseudohemiodon sp.	-	-	●	-	-
Pseudohemiodon thorectes	-	-	●	-	-
Pterosturisoma microps	-	-	●	-	-
Pterygoplichthys sp.	-	●	-	-	-
Rineloricaria beni	-	●	●	-	●
Rineloricaria lanceolata	●	●	-	-	-
Rineloricaria cf. *lanceolata*	-	-	●	-	●
Rineloricaria sp.	●	-	-	-	-
Rineloricaria sp.	-	●	-	-	-
Rineloricaria sp.	-	-	●	-	-
Scoloplax cf. *dicra*	●	-	-	-	-
Scoloplax sp.	-	-	●	-	-
Spatuloricaria cf. *evansii*	-	-	●	-	-
Sturisoma nigrirostrum	●	-	-	-	-
Sturisoma cf. *nigrirostrum*	-	-	●	-	-
Pimelodidae					
Brachyglanis? sp. **	●	-	-	-	-
Brachyplatystoma filamentosum	-	-	●	-	●
Brachyplatystoma flavicans	-	-	●	-	-
Brachyrhamdia marthae **	●	-	-	-	-
Callophysus macropterus	-	-	●	-	●
Cetopsorhamdia phantasia **	●	-	-	-	-
Cheirocerus eques **	●	-	-	-	-
Duopalatinus cf. *malarmo* **	●	-	-	-	-
Hemisorubim platyrhynchos	●	-	●	-	●
Heptapterus longior **	●	-	-	-	-
Heptapterus sp.	●	-	-	-	-
Imparfinis guttatus	-	-	●	-	-
Imparfinis sp.	●	-	-	-	-
Imparfinis stictonotus	●	-	●	-	-
Imparfinis cf. *stictonotus*	-	●	-	-	-
Leiarius marmoratus	●	-	●	-	-
Megalonema platanum	-	-	●	-	-
Megalonema sp.	●	-	-	-	-
Megalonema sp. nov.	●	-	-	-	-
Microglanis sp.	●	-	-	-	-
Microglanis sp.	-	-	●	-	-
Paulicea lutkeni	-	-	●	●	-

TAXA	Bolivia AquaRAP	Noel Kempff Mercado	Bolivian Amazon	Madre de Dios	Itenez (or Guapore)
Phenacorhamdia boliviana	-	-	●	-	-
Phractocephalus hemioliopterus	-	-	●	-	●
Pimelodella cf. *boliviana*	●	-	-	-	-
Pimelodella cf. *chaparae*	-	-	●	-	-
Pimelodella cochabambae	-	-	●	-	-
Pimelodella cristata	●	-	●	-	-
Pimelodella gracilis	●	-	●	-	-
Pimelodella hasemani **	●	-	-	-	-
Pimelodella cf. *itapicuruensis* **	●	-	-	-	-
Pimelodella mucosa	-	-	●	-	-
Pimelodella roccae	-	-	●	-	-
Pimelodella serrata	-	-	●	-	-
Pimelodella cf. *serrata*	●	-	-	-	-
Pimelodella sp.	-	-	●	-	-
Pimelodidae sp.	●	-	-	-	-
Pimelodidae sp.	-	●	-	-	-
Pimelodina flavipinnis	-	-	●	-	-
Pimelodus "*altipinnis*" **	●	-	-	-	-
Pimelodus altissimus (sp. nov.) **	●	-	-	-	-
Pimelodus armatus **	●	-	-	-	-
Pimelodus cf. *blochii*	●	-	-	-	-
Pimelodus gr. *maculatus-blochi*	-	-	●	●	●
Pimelodus ornatus	-	-	●	-	-
Pimelodus cf. *pantherinus* **	●	-	-	-	-
Pimelodus sp. 1	●	-	-	-	-
Pimelodus sp. 2	●	-	-	-	-
Pimelodus sp. 3	●	-	-	-	-
Pimelodus sp. 4	●	-	-	-	-
Pinirampus pirinampu	-	-	●	-	●
Platysilurus barbatus	-	-	●	-	-
Pseudopimelodus sp. 1	-	-	●	-	-
Pseudopimelodus sp. 2	-	-	●	-	-
Pseudoplatystoma fasciatum	●	●	●	●	●
Pseudoplatystoma tigrinum	-	-	●	●	●
Rhamdia quelen	-	-	●	-	-
Rhamdia sp.	●	-	-	-	-
Rhamdia sp.	-	●	-	-	-
Rhamdia sp.	-	-	●	-	-
Sorubim lima	●	-	●	●	●
Sorubimichthys planiceps	-	-	●	-	-
Zungaro zungaro	-	-	●	-	-
Trichomycteridae					
Acanthopoma cf. *bondi* **	●	-	-	-	-
Apomatoceros sp.	-	-	●	-	-

TAXA	Bolivia AquaRAP	Noel Kempff Mercado	Bolivian Amazon	Madre de Dios	Itenez (or Guapore)
Gyrinurus batrachostoma	-	-	●	-	-
Homodiaetus cf. *maculatus*	-	-	●	-	-
Homodiaetus sp.	●	-	-	-	-
Homodiaetus sp.	-	-	●	-	-
Ochmacanthus cf. *alternus* **	●	-	-	-	-
Ochmacanthus sp.	-	●	-	-	-
Ochmacanthus sp.	-	-	●	-	●
Paracanthopoma sp.	-	-	●	-	-
Plectrochilus sp.	●	-	-	-	-
Plectrochilus sp.	-	-	●	-	-
Pseudostegophilus nemurus	●	-	●	-	-
Trichomycteridae sp.	-	●	-	-	-
Trichomycterus cf. *barbouri*	-	-	●	-	-
Trichomycterus sp.	-	●	-	-	-
Tridentopsis pearsoni	●	-	-	-	-
Vandellia cirrhosa	●	-	●	-	-
Vandellia hasemani	-	-	●	-	-
GYMNOTIFORMES					
Apteronotidae					
Adontosternarchus clarkae **	●	-	-	-	-
Adontosternarchus sachsi	-	-	●	-	-
Apteronotus albifrons	●	-	●	-	-
Apteronotus bonapartii	●	-	●	-	-
Apteronotus sp.	-	-	●	-	●
Porotergus cf. *gimbeli*	-	-	●	-	-
Porotergus cf. *gymnotus*	-	-	●	-	-
Sternarchorhynchus oxyrhynchus	-	-	●	-	-
Sternarchorhynchus sp.	-	-	●	-	-
Electrophoridae					
Electrophorus electricus **	●	-	-	-	-
Gymnotidae					
Gymnotus anguillaris	-	-	●	-	-
Gymnotus cf. *anguillaris*	●	-	-	-	-
Gymnotus carapo	●	●	●	-	●
Gymnotus cf. *coatesi* **	●	-	-	-	-
Hypopomidae					
Brachyhypopomus brevirostris	●	-	-	-	-
Brachyhypopomus cf. *brevirostris*	-	-	●	-	-
Brachyhypopomus pinnicaudatus **	●	-	-	-	-
Brachyhypopomus sp.	●	-	-	-	-
Gymnotiforme	-	●	-	-	-
Hypopomus cf. *artedi*	-	-	●	-	-
Hypopomus sp. 1	-	●	-	-	-
Hypopomus sp. 2	-	●	-	-	-

TAXA	Bolivia AquaRAP	Noel Kempff Mercado	Bolivian Amazon	Madre de Dios	Itenez (or Guapore)
Hypopygus lepturus	●	●	-	-	-
Rhamphichthyidae					
Gymnorhamphichthys hypostomus	-	-	●	-	-
Rhamphichthys rostratus	-	●	●	●	-
Rhamphichthys sp.	-	-	●	-	-
Sternopygidae					
Distocyclus conirostris	●	-	●	-	-
Eigenmannia humboldtii	●	-	●	-	-
Eigenmannia macrops **	●	-	-	-	-
Eigenmannia cf. *trilineata* **	●	-	-	-	-
Eigenmannia virescens	●	●	●	-	●
Rhabdolichops caviceps **	●	-	-	-	-
Rhabdolichops troscheli	-	-	●	-	-
Sternopygus macrurus	●	●	●	●	●
CYPRINODONTIFORMES					
Rivulidae					
Cynolebias sp.	-	●	-	-	-
Pterolebias sp.	-	-	●	-	-
Rivulidae sp.	-	●	-	-	-
Rivulus sp.	●	-	-	-	-
Rivulus sp.	-	●	-	-	-
Rivulus sp.	-	-	●	-	-
BELONIFORMES					
Belonidae					
Potamorrhaphis cf. *eigenmanni*	-	●	-	-	-
Potamorrhaphis sp.	●	-	-	-	-
Potamorrhaphis sp.	-	-	●	-	●
Strongylura sp.	-	-	●	-	-
SYNBRANCHIFORMES					
Synbranchidae					
Synbranchus marmoratus	●	●	●	-	●
PERCIFORMES					
Cichlidae					
Aequidens sp. 1	●	-	-	-	-
Aequidens sp. 2	●	-	-	-	-
Aequidens sp. 3	●	-	-	-	-
Aequidens cf. *paraguayensis* **	●	-	-	-	-
Aequidens cf. *tetramerus*	●	-	-	-	-
Aequidens cf. *tetramerus*	-	●	-	-	-
Aequidens viridis	-	●	●	-	●
Apistogramma inconspicua	-	-	●	-	●
Apistogramma linkei	●	-	●	-	-
Apistogramma sp. 1	●	-	-	-	-
Apistogramma sp. 2	●	-	-	-	-

TAXA	Bolivia AquaRAP	Noel Kempff Mercado	Bolivian Amazon	Madre de Dios	Itenez (or Guapore)
Apistogramma sp. 3	●	-	-	-	-
Apistogramma sp. 4	●	-	-	-	-
Apistogramma sp. 1	-	●	-	-	-
Apistogramma sp. 2	-	●	-	-	-
Astronotus crassipinnis	●	●	●	-	●
Biotodoma cupido	-	●	●	-	●
Bujurquina cf. *vittata*	-	-	●	-	-
Chaetobranchiopsis orbicularis	●	-	●	●	●
Chaetobranchus flavescens	-	●	●	-	●
Cichla monoculus	-	●	●	●	●
Cichla cf. *monoculus*	●	-	-	-	-
Cichlasoma boliviense	-	-	●	-	●
Cichlasoma severum **	●	-	-	-	-
Cichlidae sp.	●	-	-	-	-
Crenicara sp.	-	-	●	-	-
Crenicara cf. *unctulata* **	●	-	-	-	-
Crenicichla cf. *heckeli* **	●	-	-	-	-
Crenicichla johanna	-	-	●	-	●
Crenicichla lepidota	-	-	●	-	●
Crenicichla semicincta	-	-	●	-	-
Crenicichla sp. 1	●	-	-	-	-
Crenicichla sp. 2	●	-	-	-	-
Crenicichla sp.	-	●	-	-	-
Crenicichla sp. A	-	-	●	-	●
Crenicichla sp. B	-	-	●	●	●
Geophagus surinamensis	-	-	●	-	●
Heros sp.	-	●	-	-	-
Heros sp.	-	-	●	●	●
Laetacara dorsiger	-	●	●	-	-
Mesonauta festivus	●	●	●	●	●
Mesonauta cf. *insignis* **	●	-	-	-	-
Microgeophagus altispinosa	●	-	●	-	●
Satanoperca cf. *acuticeps* **	●	-	-	-	-
Satanoperca jurupari	-	-	●	●	-
Satanoperca pappatera	-	-	●	-	●
Satanoperca sp.	●	-	-	-	-
Eleotridae					
Eleotridae sp.	-	-	●	-	-
Sciaenidae					
Pachypops sp.	-	-	●	-	-
Pachyurus sp.	●	-	-	-	-
Plagioscion squamosissimus	●	-	●	●	●
PLEURONECTIFORMES					
Soleidae					
Achirus achirus	-	●	●	-	-

** Species collected by AquaRAP and believed to be new records for the Bolivian Amazon.

++ Locality record from Eigenmann, C.H., and G.S. Myers. 1929. The American Characidae. Memoirs of the Museum of Comparative Zoology 43: 429-558.

Locality information for fishes in Noel Kempff Mercado is from Sarmiento (1998). Locality information for fishes in the Bolivian Amazon, Madre de Dios, and Itenez (or Guapore) columns is taken from Lauzanne et al. (1991). Note that Madre de Dios and Itenez (or Guapore) are subsections of the Bolivian Amazon, and that the Bolivian Amazon includes additional basins not indicated in this table.

Lauzanne, L., G. Loubens, and B. Le Guennec. 1991. Liste commentée des poissons de l'Amazonie bolivienne. Revista Hydrobiologia Tropical 24:61-76.

Sarmiento, J. 1998. Ichthyology of Parque Nacional Noel Kempff Mercado. *In* Killeen, T.J., and T.S. Schulenberg (eds.). A biological assessment of Parque Nacional Noel Kempff Mercado, Bolivia. Pp. 168 - 180. RAP Working Papers No. 10, Conservation International, Washington, D.C.

APPENDIX 8 Description of ichthyological field stations sampled during the AquaRAP expedition to Pando, Bolivia in September 1996.

Antonio Machado-Allison, Jaime Sarmiento, Naércio Menezes, Hernán Ortega, Soraya Barrera, Theresa M. Bert, Barry Chernoff, and Philip W. Willink

Group P1
Upper Tahuamanu Sub-Basin (Stations P01-01 to P01-10)

Field Station 96-P-01-01
Locality: Aserradero Rutina 77 km SW of Cobija.
11° 25' 55" S, 69° 00' 09" W
4/Sep/1996
Whitewater creek (2-3 m wide, 0.5 m deep), tributary of the Tahuamanu. It originates in a flooded lake. The shore and bottom are muddy and with abundant submerged logs and leaves. Water current fast. No aquatic plants. Gallery forest covering the margins. A total of 136 specimens were collected.
The species list includes:
Characiformes = 12
Siluriformes = 14
Gymnotiformes = 3
Perciformes = 1
Species total = 30
The most abundant species are *Odontostilbe paraguayensis* (41= 30.1%), *Prionobrama filigera* (16= 11.7%), and *Imparfinis stictonotus* (11= 8.1%). Other species include *Farlowella oxyrryncha* (9), *Astyanax abramis* (6), and *Eigenmannia virescens* (6). There is a predominance of fast moving and/or bottom species typical of small, fast water creeks.

Field Station 96-P-01-02
Locality: Río Tahuamanu, 2 km above Aserradero Rutina.
11° 26' 32" S, 69° 00' 42" W
4/Sep/1996
Whitewater river (70+ m wide). It originates in a flooded lake in Peru. The shore and bottom are sandy and with some submerged logs that retain leaves. The current is medium-fast. No aquatic plants. Gallery forest partially covering the margins. A total of 104 specimens were collected.
The species list includes:
Characiformes = 11
Siluriformes = 12
Species total 23
The most abundant species are *Pimelodella gracilis* (36= 34.6%), *Cheirodon fugitiva* (12= 11.5%), and *Astyanax abramis* (7= 6.7%). Other species include *Cheirocerus eques* (6), *Steindachnerina dobula* (5), and *Thoracocharax stellatus* (5). There is a predominance of fast moving and/or bottom species typical of small, fast water creeks.

Field Station 96-P-01-03

Locality: Río Tahuamanu 2/3 km above mouth of the Muyumanu.

11° 26' 21" S, 69° 02' 08" W

5/Sep/1996

Whitewater creek (50 m wide). It originates in a flooded lake in Peru. The shore and bottom are sandy/muddy and with abundant submerged logs and leaves. Water current fast. No aquatic plants. Gallery forest covering part of the margins. A total of 217 specimens were collected.

The species list includes:

Characiformes = 19

Siluriformes = 15

Perciformes = 2

Species total = 36

The most abundant species are *Prionobrama filigera* (91= 41.9%), *Pimelodella gracilis* (19= 8.7%), *Aphyocharax pusillus* (17 = 7.8%), and *Odontostilbe paraguayensis* (11= 5.0%). Other species include *Creagrutus* sp. A (9), *Astyanax abramis* (6), *Cheirodon fugitiva* (6), and *Moenkhausia dichroura* (5). There is a predominance of fast moving and/or bottom species typical of small, fast water creeks.

Field Station 96-P-01-04

Locality: Río Muyumanu, 1.5 km above mouth of Muyumanu/Tahuamanu.

Latitude and longitude unavailable.

5/Sep/1996

Whitewater creek (20 m wide). The shore and bottom are muddy and with abundant submerged logs and leaves. Water current fast. No aquatic plants. Gallery forest covering part of the margins. A total of 25 specimens were collected.

The species list includes:

Characiformes = 4

Siluriformes = 10

Species total = 14

The most abundant species are *Creagrutus* sp. A (6= 24 %), *Prionobrama filigera* (4 = 16 %), and *Imparfinis stictonotus* (3 = 12 %). Other species include *Astyanax abramis*, several pimelodids, and loricariids. There is a predominance of fast moving and/or bottom species typical of small, fast water creeks.

Field Station 96-P-01-05

Locality: Río Muyumanu, same as P1-04.

Latitude and longitude unavailable.

6/Sep/1996

The species list includes:

Characiformes = 2

Siluriformes = 1

Rajiformes = 1

Species total = 4

Collected with gillnet.

A total of 4 specimens were collected. The species are *Hypostomus* sp., *Mylossoma duriventris*, *Potamotrygon motoro*, and *Serrasalmus rhombeus*.

Field Station 96-P-01-06
Locality: Lake, flooded lake right margin of the Río Tahuamanu more or less 1000 m down river from mouth of the Río Nareuda. Latitude and longitude unavailable.
7/Sep/1996
Whitewater lake. It originates as a flooded lake, water coming from the Río Tahuamanu. The lake is used by fishermen. The shore and bottom are muddy and with abundant submerged logs and leaves. No aquatic plants. A total of 85 specimens were collected using nets.
The species list includes:
Characiformes = 21
Siluriformes = 6
Perciformes = 4
Gymnotiformes = 1
Species total = 32
The most abundant species are *Odontostilbe paraguayensis* (18 = 21.1%), *Astyanax abramis* (8 = 9.4 %), *Eigenmannia virescens* (7= 8.2 %), and *Potamorhina altamazonica*. (6 = 7.0). Other species include *Plagioscion squamosissimus, Pimelodus blochii, Aphyocharax dentatus, Odontostilbe paraguayensis, Cheirodon fugitiva, Liposarcus disjunctivus, Ctenobrycon spilurus, Apistogramma* sp., *Auchenipterus nuchalis, Hoplias malabaricus, Hydrolycus pectoralis, Pygocentrus nattereri, Prochilodus nigricans, Serrasalmus rhombeus,* and *Rhaphiodon vulpinus*. There is a predominance of slow moving predators and bottom species typical of lakes or slow water. This lake is used by fishermen. There is a note that said that in 4 hours some fishermen took about 40 kg of fish, including *Potamorhina altamazonica, Liposarcus disjunctivus, Hoplias malabaricus, Hydrolycus pectoralis, Pygocentrus nattereri, Prochilodus nigricans, Serrasalmus rhombeus,* Doradidae (*Oxydoras* ?), *Rhaphiodon vulpinus,* and *Triportheus* among others.

Field Station 96-P-01-07
Locality: Small creek on Río Muyumanu, right margin one hour from the mouth into the Tahuamanu.
11° 26' 57" S, 69° 01' 43" W
8/Sep/1996
Whitewater creek (1-2 m wide). The shore and bottom are muddy and with abundant submerged logs and leaves. The water current is medium-fast. No aquatic plants. Gallery forest covering part of the margins. A total of 124 specimens were collected.
The species list includes:
Characiformes = 12
Siluriformes = 4
Perciformes = 1
Species total = 17
The most abundant species are *Otocinclus mariae* (39 = 31.4 %), *Chrysobrycon* sp. (25 = 20.1 %), *Carnegiella myersi* (12 = 9.6 %), and *Gephyrocharax* sp. (10 = 8.0 %). Other species include *Astyanax abramis, Characidium* sp., *Gasteropelecus sternicla, Moenkhausia sanctaefilomenae,* and *Tyttocharax tambopatensis.*

Field Station 96-P-01-08
Locality: Río Muyumanu, one hour from the mouth into the Tahuamanu.
11° 26' 57" S, 69° 01' 43" W
8/Sep/1996
Whitewater creek (15 m wide). The shore and bottom are muddy. Water current medium-fast. No aquatic plants. A total of 66 specimens were collected.
The species list includes:
Characiformes = 14
Siluriformes = 15
Species total = 29
The most abundant species are *Aphanotorulus frankei* (6 = 9.0 %), *Prionobrama filigera* (6 = 9.0 %), *Astyanax abramis* (5 = 7.5 %), *Pimelodella* cf. *serrata* (4 = 6.0 %), and *Tatia altae* (4 = 6.0 %). Other species include *Odontostilbe hasemani, Phenacogaster pectinatus, Paragoniates alburnus, Sturisoma nigrirostrum, Brachychalcinus copei,* and *Moenkhausia sanctaefilomenae.*

Field Station 96-P-01-09

Locality: Small creek on Río Muyumanu, left margin half hour from the mouth into the Río Tahuamanu.
11° 27' 35" S, 69° 02' 00" W
8/Sep/1996
Whitewater creek, tea colored (1 mt wide). The shore and bottom are muddy and with abundant submerged logs and leaves. Water current medium-fast. No aquatic plants. Gallery forest covering part of the margins. A total of 14 specimens were collected.
The species list includes:
Characiformes = 2
Siluriformes = 3
Species total = 5
The most abundant species are *Characidium* sp. (6 = 43 %) and *Chrysobrycon* sp. (5 = 35.7 %). Other species include *Moenkhausia sanctaefilomenae, Otocinclus mariae,* and *Rineloricaria* sp.
Area very difficult to collect.

Field Station 96-P-01-10

Locality: Lake Canaveral. Cocha on left margin of Río Tahuamanu, 20 min. from the mouth of the Río Muyumanu.
11° 26' 15" S, 69° 01' 59" W
8/Sep/1996
Lake formed by an old arm of the river. Abundant macrophytes. Bottom and shore muddy. A total of 389 specimens were collected.
The species list includes:
Characiformes = 13
Siluriformes = 4
Gymnotiformes = 2
Perciformes = 6
Species total = 25
The most abundant species are *Odontostilbe paraguayensis* (126 = 32.3 %), *O. hasemani* (83 = 21.3 %) , *Steindachnerina dobula* (68 = 17.4 %), *Loricariichthys* sp. (21 = 5.4 %), *Aequidens* sp. B (14 = 3.6 %), *Hoplias malabaricus* (12 = 3.1 %), *Brachyrhamdia marthae* (10 = 2.6 %), and *Aequidens* sp. A (8 = 2.1 %). Other species include *Astyanax abramis, Ctenobrycon spilurus, Moenkhausia dichroura, Hypoptopoma joberti, Characidium* sp., *Moenkhausia sanctaefilomenae,* and *Crenicichla heckeli.* The species community is typical for backwaters or lagoons.

Lower Nareuda/ Middle Tahuamanu Sub-Basin (Stations P01-11 to P01-21)

Field Station 96-P-01-11

Locality: Río Nareuda 2 km above the mouth into Río Tahuamanu.
11° 18' 18" S, 68° 45' 28" W
10/Sep/1996
Blackwater river (8-10 m wide). The shore and bottom are muddy. Water current medium-fast. No aquatic plants. Gallery forest covering part of the margins. A total of 95 specimens were collected.
The species list includes:
Characiformes = 16
Siluriformes = 18
Gymnotiformes = 2
Perciformes = 2
Species total = 38
The most abundant species are *Corydoras loretoensis* (11 = 11.6 %), *Aequidens paraguayensis* (9 = 9.5 %), and *Tyttocharax madeirae* (7 = 7.4%). Other species include *Astyanax abramis, Characidium* sp., *Pimelodella gracilis, Pimelodella* sp., *Moenkhausia sanctaefilomenae,* and *Tatia altae.* The community has a predominance of species typically from blackwater rivers.

Field Station 96-P-01-12

Locality: Curichi (flooded lake) on the right margin of the Río Nareuda more or less 3-4 km from the mouth into the Río Tahuamanu.

Latitude and longitude unavailable.

10/Sep/1996

Whitewater lagoon. The shore and bottom are muddy and with abundant submerged logs and leaves. Gallery forest covering part of the margins. A total of 420 specimens were collected.

The species list include:

Characiformes = 25

Siluriformes = 14

Perciformes = 4

Species total = 43

The most abundant species are *Odontostilbe paraguayensis* (63 = 15 %), *Cyphocharax spiluropsis* (57 = 13.5 %), *Moenkhausia colletti* (45 = 10.7 %), *Phenacogaster* sp. B (41 = 9.8 %), and *Cheirodon fugitiva* (36 = 8.5 %). Other species include *Astyanax abramis, Aequidens* sp., *Brochis* sp., *Bunocephalus amazonicus, Cyphocharax* sp., *Ctenobrycon spilurus, Gasteropelecus sternicla, Moenkhausia sanctaefilomenae, Phenacogaster pectinatus, Parotocinclus* sp., *Rineloricaria lanceolata,* and *Sturisoma nigrirostrum*. The species are typically from cochas or flooded lakes.

Field Station 96-P-01-13

Locality: Río Nareuda more or less 4 km from the mouth into the Río Tahuamanu.

11° 18' 23" S, 68° 45' 57" W

10/Sep/1996

Blackwater river (12 m wide). The shore and bottom are muddy. Water current medium-fast. No aquatic plants. Gallery forest covering part of the margins. A total of 21 specimens were collected.

The species list includes:

Characiformes = 4

Siluriformes = 4

Perciformes = 1

Species total = 9

The most abundant species are *Hoplias malabaricus* (5 = 23.8 %), *Astyanax abramis* (4 = 19 %), *Rineloricaria* sp. (4 = 19 %), and *Pimelodella gracilis* (3 = 14.3 %). Other species include A*equidens paraguayensis, Imparfinis stictonotus, Moenkhausia* sp., *Prionobrama filigera,* and *Loricaria* sp.

Field Station 96-P-01-14

Locality: Río Nareuda more or less 100 m from the mouth into the Río Tahuamanu.

Latitude and longitude unavailable.

10/Sep/1996

Blackwater river (12 m wide). The shore and bottom are sandy. Water current medium-fast. No aquatic plants. Gallery forest covering part of the margins. A total of 63 specimens were collected.

The species list includes:

Characiformes = 15

Siluriformes = 11

Perciformes = 2

Rajiformes = 1

Species total = 29

The most abundant species are *Phenacogaster* sp. (7 = 11 %), *Hypoptopoma joberti* (5 = 7.9 %), *Corydoras acutus* (5 = 7.9 %), and *Bunocephalus aleuropsis* (4 = 6.3 %). Other species include A*equidens paraguayensis, Imparfinis stictonotus, Moenkhausia sanctaefilomenae, Moenkhausia* sp., *Prionobrama filigera, Pimelodella gracilis,* and *Potamotrygon motoro*. In general, these species are typically from blackwater rivers.

Field Station 96-P-01-15
Locality: Curichi (Flooded lake or dead arm), right margin of the Río Nareuda more or less 5 km from the mouth into the Río Tahuamanu.
11° 18' 32" S, 68° 45' 58" W
11/Sep/1996
Blackwater river lagoon. The shore and bottom are muddy with abundant logs and leaves. No aquatic plants. Gallery forest covering part of the margins. A total of 444 specimens were collected.
The species list includes:
Characiformes = 24
Siluriformes = 8
Perciformes = 2
Species total = 34
The most abundant species are *Cyphocharax spiluropsis* (83 = 18.7 %), *Otocinclus mariae* (61 = 13.7 %), *Charax gibbosus* (46 = 10.3 %), *Ctenobrycon spilurus* (45 = 10.1%), and *Corydoras loretoensis* (43 = 9.7 %). Other species include A*equidens tetramerus, Brachyrhamdia marthae, Bunocephalus amazonicus, Gasteropelecus sternicla, Hoplosternum thoracatus, Moenkhausia dichroura, Prochilodus nigricans, Steindachnerina dobula, Stethaprion crenatum*, and *Triportheus angulatus*. This area has a high diversity.

Field Station 96-P-01-16
Locality: Curichi (flooded lake or dead arm), right margin of the Río Nareuda more or less 5.5 km from the mouth into the Río Tahuamanu.
Latitude and longitude unavailable.
11/Sep/1996
It is a small area. The shore and bottom are muddy with lots of leaves and logs. No aquatic plants or gallery forest. A total of 5 specimens were collected.
The species list include:
Characiformes = 1
Siluriformes = 1
Species total = 2
The species are *Hoplosternum thoracatus* (3 = 60 %) and *Hoplias malabaricus* (2 = 40 %).

Field Station 96-P-01-17
Locality: Río Nareuda more or less 6 km from the mouth into the Río Tahuamanu.
11° 18' 41" S, 68° 45' 50" W
11/Sep/1996
Blackwater river (15 m wide). The shore and bottom are muddy. Water current medium-fast. No aquatic plants. Gallery forest covering part of the margins. A total of 61 specimens were collected.
The species list includes:
Characiformes = 9
Siluriformes = 7
Perciformes = 2
Species total = 18
The most abundant species are *Aequidens paraguayensis* (15 = 24.5 %), *Pimelodella gracilis* (9 = 14.7 %), *Moenkhausia* sp. (7 = 11.4 %), and *Rineloricaria* sp. (5 = 8.1 %). Other species include *Apistogramma linkei, Astyanax abramis, Corydoras loretoensis, Leporinus nattereri, Moenkhausia colletti, Ochmacanthus alternus, Prionobrama filigera,* and *Sturisoma nigrirostrum*.

Field Station 96-P-01-18
Locality: Río Nareuda (rapids) more or less 6 km from the mouth into the Río Tahuamanu.
11° 18' 18" S, 68° 45' 25" W
11/Sep/1996
Blackwater river (15 m wide). The shore and bottom are rocky. Water current fast. Algae on rocks. A total of 34 specimens were collected.
The species list includes:
Characiformes = 8
Siluriformes = 11
Species total = 19
The most abundant species are *Peckoltia arenaria* (5 = 14.7 %), *Phenacogaster pectinatus* (4 = 11.7 %), *Corydoras loretoensis* (3 = 8.8 %), and *Ancistrus* sp. (2 = 5.8%). Other species include *Astyanax abramis, Imparfinis stictonotus, Knodus victoriae, Prionobrama filigera,* and *Tatia perugiae*. Species typically from rapids.

Field Station 96-P-01-19
Locality: Small creek on the right margin (Filadelfia?).
11° 20' 33" S, 68° 46' 54" W
12/Sep/1996
Blackwater river (5 m wide). The shore and bottom are muddy with submerged logs. Water current medium. No aquatic plants. Gallery forest covering part of the margins. A total of 71 specimens were collected.
The species list includes:
Characiformes = 13
Siluriformes = 4
Gymnotiformes = 1
Perciformes = 1
Species total = 19
The most abundant species are *Phenacogaster* sp. (18 = 25.3 %), *Moenkhausia sanctaefilomenae* (9 = 12.6 %), *Otocinclus mariae* (9 = 12.6 %), *Moenkhausia colletti* (6 = 8,4 %), *Cheirodon fugitiva* (6 = 8.4 %), and *Gasteropelecus sternicla* (6 = 8.4 %). Other species include *Apistogramma* sp., *Carnegiella myersi, Corydoras acutus, Ctenobrycon spilurus, Cynopotamus gouldingi, Hemiodontichthys acipenserinus, Phenacogaster pectinatus,* and *Sorubim lima.*

Field Station 96-P-01-20
Locality: Lagoon on the right margin of the Río Tahuamanu more or less 500 m from the mouth of the Río Nareuda.
11° 18' 37" S, 68° 44' W
12/Sep/1996
Blackwater. The shore and bottom are muddy. Grasses and cyperaceans cover the margins. A total of 279 specimens were collected.
The species list includes:
Characiformes = 23
Siluriformes = 10
Perciformes = 4
Species total = 37
The most abundant species are *Corydoras loretoensis* (63 = 22.5 %), *Cyphocharax spiluropsis* (45 = 16.1 %), *Ctenobrycon spilurus* (22 = 7.8%), *Hemigrammus ocellifer* (18 = 6.5 %), and *Knodus gamma* (15 = 5.3 %). Other species include *Aequidens* sp., *Astyanax abramis, Aphanotorulus* sp., *Carnegiella myersi, Mesonauta festivus, Moenkhausia dichroura, Ochmacanthus alternus, Odontostilbe hasemani, Pimelodella gracilis, Triportheus angulatus,* and *Steindachnerina dobula.* High diversity. Species common in lagoon-like habitats.

Field Station 96-P-01-21
Locality: Río Nareuda, more or less 7 km above the mouth into the Río Tahuamanu.
11° 18' 14" S, 68° 45' 45" W
13/Sep/1996
Blackwater river (15 m wide). The shore and bottom are muddy. Water current medium-fast. No aquatic plants. Gallery forest covering part of the margins. A total of 147 specimens were collected.
The species list includes:
Characiformes = 17
Siluriformes = 13
Gymnotiformes = 3
Perciformes = 2
Species total = 35
The most abundant species are *Hypoptopoma* sp. (27 = 18.4 %), *Abramites hypselonotus* (12 = 8.1 %), *Rineloricaria lanceolata* (9 = 6.1 %), *Eigenmannia trilineata* (8 = 5.4), *Hypoptopoma joberti* (8 = 5.4 %), and *Farlowella* sp. (8 = 5.4 %). Other species include *Aequidens paraguayensis, Apteronotus albifrons, Carnegiella myersi, Cochliodon cochliodon, Creagrutus* sp., *Hoplias malabaricus, Microglanis* sp., *Myleus* sp., *Phenacogaster pectinatus,* and *Prionobrama filigera* among others.

Manuripi/Lower Tahuamanu Sub-Basin (Stations P01-22 to P01-39)

Field Station 96-P-01-22
Locality: Lake S/N 12 km from Puerto Rico above Río Manuripi.
11° 09' 14" S, 67° 33' 42" W
15/Sep/1996
Blackwater. The shore and bottom are muddy. Abundant aquatic plants (*Eichhornia, Potamogyton*, and cyperaceans). Gallery forest covering part of the margins. A total of 491 specimens were collected.
The species list includes:
Characiformes = 12
Siluriformes = 15
Gymnotiformes = 4
Perciformes = 3
Species total = 34
The most abundant species are *Corydoras loretoensis* (206 = 42 %), *Apistogramma* sp. (48 = 9.7 %), *Brachyrhamdia marthae* (44 = 8.9 %), *Hemigrammus unilineatus* (42 = 8.5 %), and *Cyphocharax spiluropsis* (30 = 6.1 %). Other species include *Acanthodoras cataphractus, Amblydoras hancockii, Brachyhypopomus* sp., *Bunocephalus amazonicus, Cheirodon piaba, Gymnotus carapo, Hemigrammus ocellifer, Hypopygus lepturus, Moenkhausia comma, M. colletti, Pyrrhulina vittata, Rineloricaria lanceolata,* and *Pimelodella gracilis* among others. This is a very diverse station.

Field Station 96-P-01-23
Locality: Río Manuripi 12 km above Puerto Rico.
11° 09' 06" S, 67° 33' 41" W
15/Sep/1996
Blackwater river (70 m wide). The shore and bottom are sandy. The water current medium-fast. Abundant cyperaceans and taropa. A total of 834 specimens were collected.
The species list includes:
Characiformes = 22
Siluriformes = 26
Gymnotiformes = 6
Perciformes = 4
Synbranchiformes = 1
Atheriniformes = 1
Species total = 60
The most abundant species are *Corydoras loretoensis* (104 = 12.4 %), *Pimelodella itapicuruensis* (75 = 9 %), *Apistogramma* sp. (75 = 9 %), *Pimelodella gracilis* (68 = 8.1 %), *Cyphocharax spiluropsis* (65 = 7.8 %), *Moenkhausia lepidura* (50 = 6 %), *Brachyhypopomus* sp. (48 = 5.8 %), *Parotocinclus* sp. (43 = 5.1 %), and *Amblydoras hancockii* (40 = 4.8 %). Other species include *Anadoras grypus, Apistogramma* sp., *Astyanax abramis, Carnegiella myersi, Corydoras acutus, Hemigrammus ocellifer, Hypoptopoma joberti, Hypopygus lepturus, Mesonauta festivus, Moenkhausia dichroura, Nannostomus trifasciatus, Ochmacanthus alternus, Prionobrama filigera, Pyrrhulina vittata, Rivulus* sp., *Sternopygus macrurus,* and *Sturisoma nigrirostrum* among others. This is a very diverse station.

Field Station 96-P-01-24
Locality: Lake (S/N) camp site, 10 km above Puerto Rico. Río Manuripi.
11° 08' 13" S, 67° 33' 41" W
15/Sep/1996
Blackwater flooded lake. The shore and bottom are muddy/sandy. Abundant aquatic plants. A total of 465 specimens were collected.
The species list includes:
Characiformes = 9
Siluriformes = 10
Gymnotiformes = 1
Perciformes = 3
Synbranchiformes = 1
Species total = 24
The most abundant species are *Apistogramma* sp. (234 = 50.3 %), *Hemigrammus lunatus* (72 = 15.5 %), *Amblydoras hancockii* (44 = 9.5 %), *Corydoras loretoensis* (44 = 9.5 %), and *Parotocinclus* sp. (15 = 3.2 %). Other species include *Astrodoras asterifrons, Brachyhypopomus* sp., *Brachyrhamdia marthae, Bunocephalus amazonicus, Cyphocharax spiluropsis, Moenkhausia colletti,* and *Synbranchus marmoratus* among others. This station possesses a high diversity of Siluriformes.

Field Station 96-P-01-25
Locality: Río Manuripi 20 km above Puerto Rico.
Latitude and longitude unavailable.
16/Sep/1996
Blackwater river (50 m wide). The shore and bottom are sandy. Water current medium-fast. Abundant aquatic plants (*Ponthederia, Eichhornia*) and cyperaceans. A total of 61 specimens were collected.
The species list includes:
Characiformes = 18
Siluriformes = 15
Gymnotiformes = 8
Perciformes = 3
Synbranchiformes = 1
Species total = 45
The most abundant species are *Pimelodella gracilis* (68 = 14.3 %), *Moenkhausia lepidura* (65 = 13.7 %), *Corydoras loretoensis* (55 = 11.6 %), *Hypoptopoma joberti* (31 = 6.5 %), *Carnegiella myersi* (24 = 5.0 %), and *Rineloricaria* sp. (23 = 4.8 %). Other species include *Apistogramma* sp., *Apteronotus albifrons, Carnegiella strigata, Cochliodon cochliodon, Corydoras acutus, Ctenobrycon spilurus, Doras eigenmanni, Eigenmannia virescens, E. macrops, Entomocorus benjamini, Hemigrammus lunatus, Hypopygus lepturus, Laemolyta* sp., *Mesonauta festivus, Moenkhausia colletti, Nannostomus trifasciatus, Sternopygus macrurus,* and *Synbranchus marmoratus* among others. The high diversity of electric fishes is quite interesting. This station has a high overall diversity.

Field Station 96-P-01-26
Locality: Río Manuripi 13 km above Puerto Rico.
Latitude and longitude unavailable.
16/Sep/1996
Blackwater river (70 m wide). The shore and bottom are sandy. Water current medium-fast. Abundant aquatic plants (*Ponthederia, Eichhornia*) and cyperaceans. A total of 555 specimens were collected.
The species list includes:
Characiformes = 12
Siluriformes = 21
Gymnotiformes = 5
Perciformes = 6
Synbranchiformes = 1
Species total = 45
The most abundant species are *Pimelodella gracilis* (84 = 15.1 %), *Corydoras loretoensis* (77 = 13.8 %), *Hemigrammus* sp. (45 = 8.1 %), *Pimelodella itapicuruensis* (43 = 7.7 %), *Cyphocharax spilurus* (37 = 6.6 %), and *Apistogramma* sp. (32 = 5.8 %). Other species include *Auchenipterichthys thoracatus, Amblydoras hancockii, Brachyhypopomus* sp., *Corydoras acutus, Eigenmannia virescens, Gasteropelecus sternicla, Hoplias malabaricus, Hypoptopoma joberti, Nannostomus trifasciatus, Mesonauta festivus, Moenkhausia colletti, Ochmacanthus alternus, Prionobrama filigera, Rineloricaria lanceolata, Sturisoma nigrirostrum,* and *Sternopygus macrurus* among others. Station with high diversity, especially of Siluriformes and Gymnotiformes.

Field Station 96-P-01-27
Locality: Río Manuripi 13 km above Puerto Rico.
Latitude and longitude unavailable.
16/Sep/1996
Blackwater river (70 m wide). The shore and bottom are sandy/muddy. Water current medium-fast. Abundant aquatic plants (*Ponthederia, Eichhornia*) and cyperaceans. A total of 577 specimens were collected.
The species list includes:
Characiformes = 13
Siluriformes = 15
Gymnotiformes = 8
Perciformes = 6
Synbranchiformes = 1
Species total = 43
The most abundant species are *Cyphocharax spiluropsis* (73 = 12.6%), *Doras eigenmanni* (59 = 10.2 %), *Apistogramma* sp. (55 = 9.5 %), *Hemigrammus lunatus* (54 = 9.5 %), *Corydoras loretoensis* (41 = 7.1 %), and *Brachyhypopomus* sp. (31 = 5.3 %). Other species include *Adontosternarchus clarkae, Amblydoras hancockii, Cheirodon piaba, Corydoras acutus, Crenicara unctulata, Eigenmannia virescens, E. humboldtii, E. trilineata, Gasteropelecus sternicla, Hoplias malabaricus, Nannostomus trifasciatus, Mesonauta festivus, Moenkhausia lepidura, Ochmacanthus alternus, Trachelyopterus* cf. *galeatus, Pimelodella gracilis, Prionobrama filigera, Rineloricaria lanceolata, Sternopygus macrurus,* and *Synbranchus marmoratus* among others. Station with high diversity, especially of Siluriformes and Gymnotiformes.

Field Station 96-P-01-28
Locality: Lake (S/N) more or less 15 km above Puerto Rico.
11° 10' 29" S, 67° 33' 52" W
17/Sep/1996
Blackwater flooded lake. The shore and bottom are sandy/muddy. Abundant aquatic plants (*Ponthederia, Eichhornia*) and cyperaceans. A total of 357 specimens were collected.
The species list includes:
Characiformes = 8
Siluriformes = 10
Gymnotiformes =2
Perciformes = 2
Synbranchiformes = 1
Species total = 23
The most abundant species are *Brachyrhamdia marthae* (137 = 38.4 %), *Hemigrammus lunatus* (97 = 27.1 %), *Apistogramma* sp. (33 = 9.2 %), and *Parotocinclus* sp. (33 = 9.2 %). Other species include *Acanthodoras cataphractus, Amblydoras hancockii, Brachyhypopomus pinnicaudatus, Corydoras napoensis, Eigenmannia trilineata, Hoplias malabaricus, Hypoptopoma joberti, Moenkhausia dichroura, Scoloplax dicra,* and *Synbranchus marmoratus* among others.

Field Station 96-P-01-29

Locality: Lake (S/N) more or less 15 km above Puerto Rico.
11° 09' 00" S, 67° 33' 37" W
17/Sep/1996
Blackwater flooded lake. The shore and bottom are sandy/muddy. Abundant aquatic plants (*Ponthederia, Eichhornia*) and cyperaceans. A total of 1014 specimens were collected.
The species list includes:
Characiformes = 11
Siluriformes = 11
Gymnotiformes = 1
Perciformes = 5
Species total = 28
The most abundant species are *Corydoras loretoensis* (479 = 47.2 %), *Brachyrhamdia marthae* (105 = 10.3 %), *Apistogramma* sp. (80 = 7.8 %), *Moenkhausia colletti* (63 = 6.2 %), *Cyphocharax* sp. (43 = 4.2 %), and *Cheirodon piaba* (36 = 3.5 %). Other species include *Amblydoras hancockii, Corydoras acutus, Crenicara unctulata, Eigenmannia trilineata, Hemigrammus unilineatus, H. ocellifer, Hoplias malabaricus, Hyphessobrycon anisitsi, Nannostomus trifasciatus, Otocinclus mariae, Pimelodella boliviana,* and *Rineloricaria lanceolata.*

Field Station 96-P-01-30

Locality: Lake (S/N) more or less 12 km above Puerto Rico.
Latitude and longitude unavailable.
17/Sep/1996
Blackwater flooded lake. The shore and bottom are sandy/muddy. Abundant aquatic plants (*Ponthederia, Eichhornia*) and cyperaceans. A total of 921 specimens were collected.
The species list includes:
Characiformes = 24
Siluriformes = 13
Perciformes =8
Species total = 45
The most abundant species are *Corydoras loretoensis* (216 = 23.4 %), *Cyphocharax spiluropsis* (182 = 19.8 %), *Apistogramma* sp. (151 = 16.4 %), *Parotocinclus* sp. (61 = 6.6 %), *Ctenobrycon spilurus* (40 = 4.3 %), and *Poptella compressa* (27 = 2.9 %). Other species include *Aequidens* sp., *Amblydoras hancockii, Brochis splendens, Corydoras acutus, Curimatella dorsalis, Cichlasoma severum, Hoplias malabaricus, Leporinus nattereri, Mesonauta festivus, Moenkhausia colletti, M. dichroura, Ochmacanthus alternus, Pimelodella gracilis, Pygocentrus nattereri, Satanoperca acuticeps,* and *Serrasalmus hollandi* among others. Station with high diversity, especially of Characiformes.

Field Station 96-P-01-31
Locality: Río Orthon more or less 2 km below Puerto Rico.
11° 05' 23" S, 67° 33' 29" W
18/Sep/1996
Whitewater river (80 m wide). The shore and bottom are sandy/muddy. Water current medium-fast. Patches of aquatic plants (*Ponthederia, Eichhornia*) and cyperaceans. A total of 332 specimens were collected.
The species list includes:
Characiformes = 23
Siluriformes = 30
Gymnotiformes = 4
Perciformes = 5
Species total = 62
The most abundant species are *Prionobrama filigera* (128 = 38.5 %), *Corydoras loretoensis* (38 = 11.4 %), *Engraulisoma taeniatum* (31 = 9.3 %), *Paragoniates alburnus* (20 = 6.0 %), and *Moenkhausia dichroura* (15 = 4.5 %). Other species include *Auchenipterichthys thoracatus, Aphyocharax dentatus, Brachyhypopomus* sp., *Brochis splendens, Corydoras acutus, C. aeneus, Crenicichla heckeli, Eigenmannia virescens, E. macrops, E. trilineata, Eucynopotamus biserialis, Hypoptopoma joberti, Knodus heterestes, Megalonema* sp. nov. (?), *Moenkhausia chrysargyrea, M. lepidura, Pachyurus* sp., *Ochmacanthus alternus, Pimelodella serrata, P. gracilis, Pimelodus blochii, Thoracocharax stellatus,* and *Tympanopleura* sp. among others. Station with very high diversity, especially of Siluriformes and Characiformes. A possible new species of *Megalonema*.

Field Station 96-P-01-32
Locality: Río Tahuamanu, 500 m above the conjunction with Río Manuripi.
Latitude and longitude unavailable.
18/Sep/1996
Whitewater river (60 m wide). The shore and bottom are sandy, with lots of logs. Water current medium-fast. A total of 40 specimens were collected.
The species list includes:
Characiformes = 8
Siluriformes = 7
Perciformes = 1
Species total = 16
The most abundant species are *Prionobrama filigera* (11 = 27.5 %), *Eucynopotamus biserialis* (6 = 15 %), *Imparfinis stictonotus* (3 = 7.5 %), and *Paragoniates alburnus* (3 = 7.5 %). Other species include *Aphanotorulus frankei, Aphyocharax dentatus, Astyanax abramis, Crossoloricaria* sp., *Galeocharax gulo, Moenkhausia lepidura, Pseudostegophilus nemurus,* and *Vandellia cirrhosa* among others.

Field Station 96-P-01-33
Locality: Lake La Anguila on Río Manuripi, more or less 1 km from the union of the Río Tahuamanu and Río Manuripi.
11° 06' 40" S, 67° 33' 20" W
18/Sep/1996
Blackwater flooded lake. The shore and bottom are sandy/muddy. Abundance of semiaquatic plants grasses and cyperaceans. A total of 631 specimens were collected.
The species list includes:
Characiformes = 13
Siluriformes = 13
Perciformes = 5
Species total = 31
The most abundant species are *Corydoras loretoensis* (212 = 33.6 %), *Poptella compressa* (127 = 20.1 %), *Moenkhausia colletti* (53 = 8.4 %), *Ctenobrycon spilurus* (40 = 6.3 %), and *Moenkhausia dichroura* (39 = 6.1 %). Other species include *Amblydoras hancockii, Ancistrus* sp., *Apistogramma* sp., *Carnegiella myersi, Corydoras acutus, Cyphocharax spiluropsis, Dianema longibarbis, Gasteropelecus sternicla,* and *Hemigrammus unilineatus* among others. This area has several species important in the aquarium trade.

Field Station 96-P-01-34

Locality: Río Manuripi, beach on the right margin 5 km from the union with the Río Tahuamanu.

11° 07' 38" S, 67° 33' 29" W

18/Sep/1996

Blackwater river (100 m wide). The shore and bottom are muddy. Water current slow. Patches of aquatic vegetation (*Ponthederia, Eichhornia*), grasses, and cyperaceans on the margins. A total of 639 specimens were collected.

The species list includes:

Characiformes = 17

Siluriformes = 28

Gymnotiformes = 5

Perciformes = 3

Species total = 53

The most abundant species are *Eigenmannia macrops* (222 = 34.7 %), *Doras* cf. *carinatus* (59 = 9.2 %), *Creagrutus* sp. (46 = 7.1 %), *Moenkhausia colletti* (39 = 6.1 %), *Corydoras loretoensis* (38 = 5.9 %), and *Apistogramma* sp. (26 = 4.1 %). Other species include *Auchenipterichthys thoracatus, Amblydoras hancockii, Brochis splendens, Corydoras acutus, Eigenmannia virescens, E. humboldtii, Entomocorus benjamini, Gasteropelecus sternicla, Hemidoras microstomus Hoplias malabaricus, Megalonema* sp. nov., *Mesonauta festivus, Moenkhausia megalops, Ochmacanthus alternus, Opsodoras humeralis, Pimelodella gracilis, Prionobrama filigera, Rineloricaria lanceolata, Serrasalmus hollandi,* and *Trachydoras paraguayensis* among others. Station with high diversity, especially of Siluriformes and Characiformes. High density of electric fishes. Several species very important in the aquarium trade.

Field Station 96-P-01-35

Locality: Río Manuripi, arm at 1 km above base camp.

11° 08' 32" S, 67° 33' 33" W

18/Sep/1996

Blackwater river (30 - 40 m wide). The shore and bottom are sandy/muddy. Water current medium-fast. Abundant aquatic plants (*Ponthederia, Eichhornia*), grasses, and cyperaceans on the margins. A total of 335 specimens were collected.

The species list includes:

Characiformes = 20

Siluriformes = 13

Gymnotiformes = 2

Perciformes = 2

Species total = 37

The most abundant species are *Cyphocharax spiluropsis* (51 = 15.2 %), *Pimelodella gracilis* (51 = 15.2 %), *Corydoras loretoensis* (37 = 11.0 %), *Pimelodella itapicuruensis* (28 = 8.3 %), *Eigenmannia macrops* (25 = 7.4 %), and *Ctenobrycon spilurus* (18 = 5.3 %). Other species include *Apistogramma* sp., *Corydoras acutus, Curimatella dorsalis, Eigenmannia virescens, Hemiodontichthys acipenserinus, Hypoptopoma joberti, Nannostomus trifasciatus, Moenkhausia colletti, M. lepidura, Ochmacanthus alternus, Phenacogaster microstictus, P. pectinatus, Pimelodella cristata, Prionobrama filigera, Rineloricaria* sp., and *Triportheus angulatus* among others. Station with medium diversity, especially of Characiformes and Siluriformes. Several species very important in the aquarium trade.

Field Station 96-P-01-36

Locality: Lake La Anguila on Río Manuripi, more or less 1 km from the union of the Río Tahuamanu and Río Manuripi (same as P1-33).

11° 06' 40" S, 67° 33' 20" W

19/Sep/1996

Blackwater flooded lake. The shore and bottom are sandy/muddy. Abundant semiaquatic plants, grasses, and cyperaceans. A total of 127 specimens were collected.

The species list includes:
Characiformes = 10
Siluriformes = 6
Perciformes = 2
Species total = 18

The most abundant species are *Psectrogaster curviventris* (23 = 18.1 %), *Potamorhina laitior* (22 = 17.3 %), *Hemigrammus lunatus* (22 = 17.3 %), *Poptella compressa* (21 = 16.5 %), and *Glyptoperichthys lituratus* (6 = 4.7 %). Other species include *Cichla monoculus, Cyphocharax spiluropsis, Liposarcus disjunctivus, Platydoras costatus, Potamorhina altamazonica, Pseudoplatystoma fasciatum, Serrasalmus hollandi,* and *Triportheus angulatus* among others. Station with medium diversity. Some species of commercial importance in aquarium trade. (Atarraya).

Field Station 96-P-01-37

Locality: Lake (S/N) on Río Manuripi, 9 km from Puerto Rico.

11° 07' 59" S, 67° 33' 28" W

20/Sep/1996

Blackwater flooded lake. The shore and bottom are sandy/muddy. Riparian forest on margins. A total of 893 specimens were collected.

The species list includes:
Characiformes = 20
Siluriformes = 13
Perciformes = 6
Species total = 39

The most abundant species are *Poptella compressa* (200 = 22.4 %), *Moenkhausia colletti* (181 = 20.2 %), *Apistogramma* sp. (96 = 10.8 %), *Corydoras loretoensis* (76 = 8.5 %), and *Cyphocharax spiluropsis* (45 = 5.0 %). Other species include *Auchenipterus thoracatus, Agamyxis pectinifrons, Amblydoras hancockii, Brachyrhamdia marthae, Corydoras acutus, Ctenobrycon spilurus, Gasteropelecus sternicla, Hemigrammus unilineatus, Moenkhausia chrysargyrea, Trachelyopterus* cf. *galeatus, Satanoperca acuticeps,* and *Tatia aulopigia* among others. Medium/high diversity. Several abundant species of commercial importance in aquarium trade.

Field Station 96-P-01-38
Locality: Río Manuripi, 8 km above Puerto Rico.
11° 07' 32" S, 67° 33' 25" W
20/Sep/1996
Blackwater river. The shore and bottom are sandy. Some aquatic and semiaquatic plants (*Eichhornia* and *Ponthederia*), grasses, and cyperaceans. A total of 153 specimens were collected.
The species list includes:
Characiformes = 7
Siluriformes = 13
Gymnotiformes = 2
Perciformes = 3
Species total = 25
The most abundant species are Corydoras loretoensis *(51 = 33.3 %),* Parotocinclus *sp. (24 = 15.7 %), Knodus caquetae* (9 = 5.8 %), and *Apistogramma* sp. (7 = 4.6 %). Other species include *Apteronotus albifrons, Brachyrhamdia marthae, Corydoras acutus, Crenicichla heckeli, Ctenobrycon spilurus, Eigenmannia virescens, Moenkhausia colletti, Pimelodella cristata*, and *Rineloricaria* sp. among others. Medium diversity. Several species of commercial importance in aquarium trade.

Field Station 96-P-01-39
Locality: Lake La Anguila on Río Manuripi, more or less 1 km from the union of the Río Tahuamanu and Río Manuripi (same as P1-33).
11° 06' 40" S, 67° 33' 20" W
20/Sep/1996
Blackwater flooded lake. The shore and bottom are sandy/muddy. Abundant semiaquatic plants, grasses, and cyperaceans. A total of 64 specimens were collected.
The species list includes:
Characiformes = 7
Siluriformes = 1
Species total = 8
The most abundant species are *Psectrogaster curviventris* (32 = 50 %), *Potamorhina laitior* (12 = 18.8 %), and *Engraulisoma taeniatum* (9 = 14 %). Other species include *Metynnis luna, Poptella compressa,* and *Triportheus angulatus* among others. Station with low diversity. Some species of commercial importance for human consumption. Collected with Tarrafa (Atarraya).

Group P2
Upper Nareuda Sub-Basin (Stations P02-01 to P02-13)

Field Station 96-P-02-01
Locality: Río Nareuda, above camp Nareuda (at beach).
11° 16' S, 69° 04' W
4/Sep/1996
White turbid water river. The shore and bottom are sandy/muddy. No semiaquatic or aquatic plants. Water current moderate. A total of 80 specimens were collected.
The species list includes:
Characiformes = 6
Siluriformes = 13
Perciformes = 2
Species total = 21
The most abundant species are *Knodus gamma* (24 = 30 %), *Pimelodella gracilis* (19 = 23.7 %), and *Hyphessobrycon gracilior* (8 = 10 %). Other species include *Aequidens paraguayensis, Brachychalcinus copei, Bunocephalus amazonicus, Corydoras acutus, Crenicichla heckeli, Homodiaetus* sp., *Pseudocetopsis* sp., *Rineloricaria lanceolata*, and *Vandellia cirrhosa* among others. Station with medium diversity.

Field Station 96-P-02-02
Locality: Río Nareuda at camp.
11° 16' S, 69° 04' W
4/Sep/1996
White turbid water river (8 m wide). The shore and bottom are sandy/muddy with some submerged rocks and logs. No aquatic plants. Gallery forest. A total of 425 specimens were collected.
The species list includes:
Characiformes = 22
Siluriformes = 9
Perciformes = 2
Species total = 33
The most abundant species are *Odontostilbe hasemani* (225 = 53 %), *Aphyocharax dentatus* (37 = 8.7 %), *Phenacogaster* sp. (27 = 6.3 %), *Bryconamericus* sp. (24 = 5.6 %), *Knodus gamma* (17 = 4 %), and *Creagrutus* sp. (14 = 3.2 %). Other species include *Aequidens paraguayensis, Aphanotorulus frankei, Aphyocharax alburnus, Brachychalcinus copei, Characidium* sp., *Gasteropelecus sternicla, Megalonema* sp., *Moenkhausia sanctaefilomenae, Otocinclus mariae, Pimelodella gracilis, Prionobrama filigera,* and *Steindachnerina dobula* among others. Station with medium diversity. Some species of importance in aquarium trade.

Field Station 96-P-02-03
Locality: Río Nareuda, below bridge covered about 300 yards.
11° 16' 39" S, 69° 03' 57" W
4/Sep/1996
White turbid water river (8 m wide). The shore and bottom are sandy/muddy with some submerged rocks and logs. No aquatic plants. Gallery forest. A total of 52 specimens were collected.
The species list includes:
Characiformes = 8
Siluriformes = 13
Perciformes = 2
Beloniformes = 1
Species total = 24
The most abundant species are *Hypoptopoma* sp. (7 = 13.4 %), *Rineloricaria lanceolata* (5 = 9.6 %), *Sturisoma nigrirostrum* (4 = 7.6 %), *Moenkhausia sanctaefilomenae* (4 = 7.6 %), and *Moenkhausia* sp. (4 = 7.6 %). Other species include *Aequidens* sp., *Ancistrus* sp., *Apistogramma* sp., *Characidium* sp., *Cochliodon cochliodon, Corydoras acutus, Farlowella* sp., *Pimelodella gracilis, Pseudocetopsis* sp., *Potamorrhaphis* sp., *Sturisoma nigrirostrum, Tatia altae,* and *Tyttocharax madeirae* among others. Station with low diversity. Some species of commercial value.

Field Station 96-P-02-04
Locality: Río Nareuda just above camp.
Latitude and longitude unavailable.
4/Sep/1996
White turbid water river (8 m wide). The shore and bottom are sandy/muddy with some rocks and logs submerged. No aquatic plants. Gallery forest. A total of 38 specimens were collected.
The species list includes:
Characiformes = 10
Siluriformes = 6
Species total = 16
The most abundant species are *Mylossoma duriventris* (9 = 23.7 %), *Hydrolycus pectoralis* (5 = 13.1 %), *Cochliodon cochliodon* (3 = 7.9 %), *Prochilodus nigricans* (3 = 7.9 %), and *Rhaphiodon vulpinus* (2 = 5.2 %). Other species include *Ageneiosus* sp., *Hemisorubim platyrhynchus, Leporinus friderici, Pimelodella cristata, Pimelodus armatus, Schizodon fasciatus, Serrasalmus rhombeus,* and *Triportheus angulatus* among others. Collected with gillnets.

Field Station 96-P-02-05

Locality: Río Nareuda 1 hour above camp, by caño coming from the forest.

Latitude and longitude unavailable.

5/Sep/1996

White turbid water small river. The shore and bottom are sandy/muddy with some rocks and logs submerged. No aquatic plants. Gallery forest. A total of 119 specimens were collected.

The species list includes:

Characiformes = 14

Siluriformes = 11

Gymnotiformes = 1

Perciformes = 4

Synbranchiformes = 1

Species total = 18

The most abundant species are *Knodus gamma* (21 = 17.6 %), *Phenacogaster pectinatus* (18 = 15.1 %), *Bryconamericus* cf. *peruanus* (17 = 14.3 %), *Aphanotorulus frankei* (14 = 11.7 %), and *Aequidens paraguayensis* (5 = 4.2 %). Other species include *Apistogramma* spp., *Bunocephalus depressus, Carnegiella myersi, Corydoras acutus, Gasteropelecus sternicla, Gymnotus coatesi, Homodiaetus* sp., *Imparfinis stictonotus, Otocinclus mariae, Steindachnerina dobula,* and *Synbranchus marmoratus* among others. Station with low diversity.

Field Station 96-P-02-06

Locality: Río Nareuda by caño coming from the forest.

Latitude and longitude unavailable.

5/Sep/1996

White turbid water small river. The shore and bottom are sandy/muddy with some gravel and submerged leaves. No aquatic plants. Gallery forest. A total of 71 specimens were collected.

The species list includes:

Characiformes = 4

Siluriformes = 9

Gymnotiformes = 1

Perciformes = 1

Species total = 15

The most abundant species are *Characidium* sp. (17 = 23.9 %), *Otocinclus mariae* (16 = 22.5 %), *Bryconamericus* cf. *peruanus* (10 = 14.1 %), and *Microglanis* sp. (5 = 7 %). Other species include *Ancistrus* sp., *Apistogramma* sp., *Cochliodon cochliodon, Chrysobrycon* sp., *Moenkhausia sanctaefilomenae,* and *Phenacogaster pectinatus* among others. Station with low diversity, however was an interesting locality because many species collected here, such as *Microglanis,* were not collected in the Nareuda proper.

Field Station 96-P-02-07
Locality: Río Nareuda 1 km just below caño coming from the forest.
Latitude and longitude unavailable.
5/Sep/1996
White turbid water small river. The shore and bottom are sandy/muddy with some gravel and submerged leaves. No aquatic plants. Gallery forest. A total of 43 specimens were collected.
The species list includes:
Characiformes = 9
Siluriformes = 10
Gymnotiformes = 1
Perciformes = 2
Synbranchiformes = 1
Species total = 23
The most abundant species are *Pimelodella gracilis* (6 = 13.9 %), *Creagrutus* sp. (5 = 11.6 %), *Hypoptopoma* sp. (4 = 9.3 %), and *Hypostomus* sp. (4 = 9.3 %). Other species include *Aequidens paraguayensis, Bunocephalus* sp., *Cochliodon cochliodon, Corydoras acutus, Galeocharax gulo, Gymnotus anguillaris, Steindachnerina dobula,* and *Synbranchus marmoratus.*

Field Station 96-P-02-08
Locality: Río Nareuda at beach 200 m below the bridge.
11° 16' 39" S, 69° 03' 57" W
5/Sep/1996
White turbid water small river. The shore and bottom are sandy/muddy with some gravel and submerged leaves. No aquatic plants. Gallery forest. A total of 57 specimens were collected.
The species list includes:
Characiformes = 8
Siluriformes = 9
Perciformes = 2
Species total = 19
The most abundant species are *Knodus gamma* (14 = 24.6 %), *Pimelodella gracilis* (7 = 12.3 %), *Knodus* sp. (6 = 10.5 %), and *Aequidens paraguayensis* (5 = 8.8 %). Other species include *Ancistrus* sp., *Corydoras acutus, Crenicichla heckeli, Imparfinis stictonotus, Moenkhausia colletti, Phenacogaster* sp., and *Tyttocharax madeirae* among others. Station with low diversity.

Field Station 96-P-02-09
Locality: Río Tahuamanu, small river at bridge on road to Cobija.
11° 14' 29" S, 68° 59' 33" W
7/Sep/1996
Small black, but turbid, water igarape (less than 3 m wide). The shore and bottom are sandy/muddy with some submerged leaves and logs. Some aquatic plants. Gallery forest. A total of 157 specimens were collected.
The species list includes:
Characiformes = 14
Siluriformes = 5
Gymnotiformes = 1
Species total = 20
The most abundant species are *Moenkhausia colletti* (42 = 26.7 %), *Bryconamericus peruanus* (32 = 20.3 %), *Phenacogaster pectinatus* (16 = 10.2 %), and *Tyttocharax tambopatensis* (15 = 9.6 %). Other species include *Brachychalcinus copei, Corydoras trilineatus, Eigenmannia virescens, Farlowella* sp., *Pyrrhulina vittata, Rineloricaria lanceolata,* and *Steindachnerina guentheri.*

Field Station 96-P-02-10
Locality: Río Nareuda 1 hour above camp, by caño coming from the forest.
11° 16' 33" S, 69° 04' 30" W
7/Sep/1996
White turbid water small river (1.5 m). The shore and bottom are sandy/muddy with some submerged sticks and leaves. No aquatic plants. Gallery forest. A total of 34 specimens were collected.
The species list includes:
Characiformes = 6
Siluriformes = 3
Perciformes = 3
Species total = 12
The most abundant species are *Pimelodella gracilis* (8 = 23.5 %), *Carnegiella myersi* (6 = 17.6 %), *Phenacogaster pectinatus* (5 = 14.7 %), and *Moenkhausia sanctaefilomenae* (4 = 11.7%). Other species include *Aequidens paraguayensis, Apistogramma linkei, Corydoras loretoensis, C. acutus, Crenicichla heckeli, Cynopotamus gouldingi,* and *Moenkhausia colletti* among others. Low diversity.

Field Station 96-P-02-11
Locality: Río Nareuda, caño coming from the forest.
11° 16' 33" S, 69° 04' 31" W
7/Sep/1996
White turbid water small river (2 m). The shore and bottom are sandy/muddy with some gravel and submerged leaves. No aquatic plants. Gallery forest. A total of 25 specimens were collected.
The species list includes:
Characiformes = 7
Siluriformes = 3
Gymnotiformes = 1
Perciformes = 1
Species total = 12
The most abundant species are *Eigenmannia macrops* (10 = 23.9 %), *Moenkhausia colletti* (2 = 22.5 %), *Phenacogaster pectinatus* (2 = 14.1 %), and *Tyttocharax* sp. nov. (2 = 7 %). Other species include *Ancistrus* sp., *Apistogramma* sp., *Cochliodon cochliodon, Chrysobrycon* sp., *Moenkhausia sanctaefilomenae,* and *Phenacogaster pectinatus* among others. Station with low diversity, however was an interesting locality because many species collected here, such as *Microglanis*, were not collected in the Nareuda proper.

Field Station 96-P-02-12
Locality: Río Nareuda by caño coming from the forest.
11° 17' 27" S, 69° 04' 41" W
8/Sep/1996
White turbid water small river. The shore and bottom are sandy/muddy with some submerged logs and leaves. Some aquatic plants. Gallery forest. A total of 15 specimens were collected.
The species list includes:
Characiformes = 4
Siluriformes = 3
Perciformes = 1
Species total = 8
The most abundant species are *Chrysobrycon* sp. (4 = 26.6 %), *Moenkhausia colletti* (3 = 20 %), *Bunocephalus amazonicus* (2 =13.3 %), and *Tyttocharax madeirae* (2 = 13.3 %). Other species include *Aphyocharax dentatus, Crenicichla* sp., *Imparfinis stictonotus,* and *Tyttocharax tambopatensis* among others. Station with very low diversity.

Field Station 96-P-02-13
Locality: Garape Campo Franza
11° 17' 06" S, 69° 04' 24" W
8/Sep/1996
White turbid water small river. The shore and bottom are sandy/muddy, with some submerged logs and leaves. No aquatic plants. Gallery forest disturbed by cattle ranching. A total of 93 specimens were collected.
The species list includes:
Characiformes = 15
Siluriformes = 8
Gymnotiformes = 1
Perciformes = 2
Species total = 26
The most abundant species are *Tyttocharax tambopatensis* (18 = 19.4 %), *Moenkhausia colletti* (17 = 18.3 %), *Pyrrhulina vittata* (7 = 7.5 %), *Gasteropelecus sternicla* (7 = 7.5 %) and *Apistogramma* sp. (5 = 5.3 %). Other species include *Ancistrus* sp., *Apistogramma* sp., *Cochliodon cochliodon, Chrysobrycon* sp., *Moenkhausia sanctaefilomenae,* and *Phenacogaster pectinatus* among others. Station with low diversity, however was an interesting locality because many species collected here, such as *Microglanis,* were not collected in the Nareuda proper.

Middle Tahuamanu Sub-Basin (Stations P02-14 to P02-27)

Field Station 96-P-02-14
Locality: Río Tahuamanu, 15 min. from the mouth of the Río Nareuda.
11° 17' 39" S, 68° 44' 23" W
10/Sep/1996
White turbid water river (100 m wide). The shore and bottom are sandy/muddy. No aquatic plants. Gallery forest. A total of 100 specimens were collected.
The species list includes:
Characiformes = 12
Siluriformes = 7
Perciformes = 1
Species total = 20
The most abundant species are *Crenicichla heckeli* (29 = 29 %), *Pimelodella itapicuruensis* (16 = 16 %), *Prionobrama filigera* (13 = 13 %), *Acanthopoma bondi* (8 = 8%), and *Aphyocharax dentatus* (6 = 6 %). Other species include *Aphanotorulus frankei, Astyanax abramis, Centromochlus heckeli, Farlowella* sp., *Moenkhausia dichroura, Pimelodella hasemani,* and *Thoracocharax stellatus* among others. Station with low diversity.

Field Station 96-P-02-15
Locality: Río Tahuamanu at sand island across lake, 1.93 km below Río Nareuda mouth (same as P02-27).
11° 17' 33" S, 68° 44' 28" W
10/Sep/1996
White turbid water river (100 m wide). The shore and bottom are sandy/muddy. No aquatic plants. A total of 40 specimens were collected.
The species list includes:
Characiformes = 7
Siluriformes = 5
Perciformes = 1
Species total = 13
The most abundant species are *Apistogramma* sp. (12 = 30 %), *Pimelodella gracilis* (5 = 12.5 %), *Moenkhausia* sp. (4 = 10 %), and *Crossoloricaria* sp. (4 = 10 %). Other species include *Aphyocharax dentatus, Clupeacharax anchoveoides, Engraulisoma taeniatum, Galeocharax gulo, Peckoltia arenaria,* and *Steindachnerina* sp. among others. Station with low diversity.

Field Station 96-P-02-16

Locality: Río Tahuamanu below camp, 0.99 km below Río Nareuda mouth.

11° 16' 24" S, 68° 44' 13" W

10/Sep/1996

White turbid water river (100 m wide). The shore and bottom are sandy/muddy with some submerged logs and leaves. No aquatic plants. A total of 211 specimens were collected.

The species list includes:

Characiformes = 17

Siluriformes = 10

Perciformes = 1

Clupeiformes = 1

Species total = 29

The most abundant species are *Odontostilbe hasemani* (53 = 25.1 %), *Aphanotorulus frankei* (49 = 23.2 %), *Knodus* sp. (23 = 10.9 %), *Pimelodella itapicuruensis* (13 = 6.1 %), and *Anchoviella carrikeri* (11 = 5.2 %). Other species include *Abramites hypselonotus, Acanthopoma bondi, Creagrutus* sp., *Moenkhausia dichroura, Paragoniates alburnus*, and *Prionobrama filigera* among others. Station with medium diversity, however was a good collection of *Aphanotorulus* and *Anchoviella*.

Field Station 96-P-02-17

Locality: Río Tahuamanu, below camp along sandy beaches.

11° 16' 22" S, 68° 44' 16" W

10/Sep/1996

White turbid water river (100 m wide). The shore and bottom are sandy/muddy. No aquatic plants. A total of 77 specimens were collected.

The species list includes:

Characiformes = 3

Siluriformes = 8

Species total = 11

The most abundant species are *Megalonema* sp. nov. (35 = 45.4 %), *Creagrutus* sp. (24 = 31.2 %), and *Pimelodus* sp. (11 = 5.2 %). Other species include *Cetopsorhamdia phantasia, Engraulisoma taeniatum, Phenacogaster* sp., *Planiloricaria cryptodon, Pseudohemiodon* sp., and *Vandellia cirrhosa* among others. Station with low diversity, however was a good collection of a new species of *Megalonema*.

Field Station 96-P-02-18

Locality: Garape Preto, ca 300 m above mouth into Río Tahuamanu, 4.36 km below mouth of Río Nareuda.

Latitude and longitude unavailable.

11/Sep/1996

Blackwater small river (5-6 m wide). The shore and bottom are sandy/muddy with some submerged logs and leaves. No aquatic plants. Dense gallery forest. A total of 84 specimens were collected.

The species list includes:

Characiformes = 10

Siluriformes = 7

Gymnotiformes = 1

Perciformes = 2

Beloniformes = 1

Species total = 21

The most abundant species are *Moenkhausia colletti* (30 = 35.7 %), *Apistogramma* sp. (10 = 11.9 %), *Moenkhausia sanctaefilomenae* (9 = 10.7 %), and *Otocinclus mariae* (7 = 8.3 %). Other species include *Aequidens paraguayensis, Carnegiella myersi, Cochliodon cochliodon, Cyphocharax spiluropsis, Corydoras loretoensis, Moenkhausia lepidura, Potamorrhaphis* sp., *Pyrrhulina vittata*, and *Tyttocharax madeirae* among others. Station with medium diversity.

Field Station 96-P-02-19

Locality: Garape Preto, above mouth at Chachalita (?) in Río Tahuamanu, 4.36 km below mouth of Río Nareuda.
11° 16' 21" S, 68° 44' 15" W
11/Sep/1996
Blackwater water river (4 m wide). The shore highly disturbed. Shore and bottom are sandy/muddy with some submerged logs and leaves. Some rooted aquatic plants and grasses. A total of 74 specimens were collected.
The species list includes:
Characiformes = 9
Siluriformes = 6
Perciformes = 1
Species total = 16
The most abundant species are *Moenkhausia colletti* (24 = 32.4 %), *Ochmacanthus alternus* (9 = 12.1 %), *Moenkhausia sanctaefilomenae* (6 = 8.1 %), and *Otocinclus mariae* (5 = 6.8 %). Other species include *Apistogramma* sp., *Characidium* sp., *Cochliodon cochliodon, Farlowella oxyrryncha, Microschemobrycon geisleri(*), Phenacogaster* sp., and *Pimelodella gracilis* among others. Station with low diversity, however was first station with *Microschemobrycon.*

Field Station 96-P-02-20

Locality: Río Tahuamanu at large sandy spit and beach across river on muddy shore below Cachmelita (?).
11° 16' 11" S, 68° 43' 55" W
11/Sep/1996
White turbid water river (100 m wide). The shore and bottom are sandy/muddy. No aquatic plants. A total of 23 specimens were collected.
The species list includes:
Characiformes = 1
Siluriformes = 2
Species total = 3
The most abundant species are *Aphanotorulus frankei* (15 = 65.2 %), *Steindachnerina* sp. (7 = 30.4 %), and *Pimelodus blochii* (1 = 4.3 %). Station with only 3 species collected.

Field Station 96-P-02-21

Locality: Río Tahuamanu below mouth of Nareuda.
11° 16' 22" S, 68° 44' 16" W
11/Sep/1996
White turbid water river (100 m wide). The shore and bottom are sandy/muddy with some submerged logs and leaves. No aquatic plants. A total of 21 specimens were collected.
The species list includes:
Characiformes = 2
Siluriformes = 5
Species total = 7
The most abundant species are *Megalonema* sp. nov. (9 = 42.8 %) and *Creagrutus* sp. (7 = 33.3 %). Other species include *Odontostilbe eques, Cochliodon cochliodon, Cheirodon fugitiva, Pimelodella gracilis,* and *Pimelodella itapicuruensis.* Station with very low diversity, however we collected several specimens of the new species of *Megalonema.*

Field Station 96-P-02-22
Locality: Río Tahuamanu at rocky island archipelago and rapids, 6.8 km below mouth of Río Nareuda.
11° 18' 09" S, 68° 44' 28" W
12/Sep/1996
White turbid water river (100 m wide). The shore and bottom are rocky, sandy/muddy with some logs. No aquatic plants. A total of 185 specimens were collected.
The species list includes:
Characiformes = 18
Siluriformes = 12
Gymnotiformes = 1
Species total = 31
The most abundant species are *Odontostilbe hasemani* (57 = 30.8 %), *Pimelodella gracilis* (17 = 9.1 %), *Aphyocharax dentatus* (12 = 6.4 %), *Prionobrama filigera* (11 = 5.9 %), and *Knodus victoriae* (10 = 5.4 %). Other species include *Abramites hypselonotus, Aphanotorulus frankei Clupeacharax anchoveoides, Moenkhausia dichroura, Paragoniates alburnus, Pimelodella serrata, Rineloricaria lanceolata Roeboides* sp., and *Thoracocharax stellatus* among others. Station with medium to high diversity. Fishes typical of rapids.

Field Station 96-P-02-23
Locality: Río Tahuamanu, small rapids just above mouth of Río Nareuda.
11° 18' 51" S, 68° 44' 35" W
12/Sep/1996
White turbid water river (100 m wide). The shore and bottom are rocky, sandy/muddy. No aquatic plants. A total of 32 specimens were collected.
The species list includes:
Characiformes = 5
Siluriformes = 10
Gymnotiformes = 1
Species total = 16
The most abundant species are *Knodus gamma* (5 = 15.6 %), *Abramites hypselonotus* (4 = 12.5 %), *Pimelodella gracilis* (4 = 12.5 %), *Odontostilbe paraguayensis* (3 = 9.0 %), and *Pimelodus pantherinus* (2 = 5.8 %). Other species include *Aphanotorulus frankei, Eigenmannia virescens, Galeocharax gulo, Leiarius marmoratus, Panaque* sp., *Peckoltia arenaria,* and *Prionobrama filigera* among others. Station with low diversity, however included first records of some species such as *Panaque* and *Leiarius marmoratus.*

Field Station 96-P-02-24
Locality: Lake with canal off Río Tahuamanu, 1.93 km below mouth of Río Nareuda.
11° 17' 32" S, 68° 44' 35" W
12/Sep/1996
White turbid water flooded lake. The shore and bottom are muddy with some submerged terrestrial plants. No aquatic plants. A total of 95 specimens were collected.
The species list includes:
Characiformes = 14
Siluriformes = 3
Perciformes = 3
Species total = 20
The most abundant species are *Odontostilbe paraguayensis* (20 = 21.1 %), *Aequidens paraguayensis* (14 = 14.7 %), *Hemigrammus lunatus* (13 = 13.6 %), *Odontostilbe hasemani* (13 = 13.6 %), and *Cyphocharax spiluropsis* (10 = 10.5 %). Other species include *Aphanotorulus frankei, Crenicichla heckeli, Gasteropelecus sternicla, Moenkhausia colletti, M. dichroura,* and *Prionobrama filigera* among others. Station with low diversity.

Field Station 96-P-02-25
Locality: Small arroyo leaving the forest just below mouth of Río Nareuda.
11° 18' 32" S, 68° 44' 21" W
12/Sep/1996
White water creek (0.5 m wide). The shore and bottom are sandy/muddy with some submerged logs and leaves. No aquatic plants.
A total of 13 specimens were collected.
The species list includes:
Characiformes = 2
Siluriformes = 5
Species total = 7
The most abundant species are *Doras* cf. *carinatus* (6 = 46.2 %) and *Otocinclus mariae* (2 = 15:3 %). Other species include *Bunocephalus aleuropsis, Characidium* sp., *Farlowella* sp., *Hoplias malabaricus, Imparfinis stictonotus,* and *Prionobrama filigera* among others. Station with very low diversity, however was a good collection of *Doras* cf. *carinatus.*

Field Station 96-P-02-26
Locality: Río Tahuamanu from mouth of Río Nareuda to below Cachuelita.
Latitude and longitude unavailable.
13/Sep/1996
White turbid water river (100 m wide). The shore and bottom are sandy/muddy with some submerged logs and leaves. A total of 258 specimens were collected.
The species list includes:
Characiformes = 6
Siluriformes = 18
Gymnotiformes = 1
Other = 1
Species total = 29
The most abundant species are *Megalonema* sp. nov. (116 = 44.9 %), *Creagrutus* sp. A (31 = 12 %), *Pimelodus "altipinnis"* (28 = 10.8 %), *Loricaria* sp. (15 = 5.8%), and *Creagrutus* sp. B (10 = 3.8 %). Other species include *Crossoloricaria* sp., *Eigenmannia macrops, E. virescens, Lamontichthys filamentosus, Opsodoras stubelii, Panaque* sp., *Plectrochilus* sp., *Trachydoras atripes,* and *Xiliphius melanopterus* among others. This is an interesting collection using trawls. Several new records were found in this area. (Trawl)

Field Station 96-P-02-27
Locality: Río Tahuamanu at sand island 1.93 km below mouth of Río Nareuda (same as P02-15).
11° 17' 33" S, 68° 44' 28" W
13/Sep/1996
White turbid water river (100 m wide). The shore and bottom are muddy. No aquatic plants. A total of 379 specimens were collected.
The species list includes:
Characiformes = 15
Siluriformes = 10
Gymnotiformes = 2
Species total = 27
The most abundant species are *Aphyocharax dentatus* (196 = 51.7 %), *Moenkhausia dichroura* (50 = 13.1 %), *Aphanotorulus frankei* (28 = 7.3 %), *Pimelodella gracilis* (13 = 3.4 %), and *Pimelodella itapicuruensis* (11 = 2.9 %). Other species include *Cheirocerus eques, Clupeacharax anchoveoides, Eigenmannia macrops, E. virescens, Farlowella* sp., *Knodus* spp. (4), *Megalonema* sp. nov., *Odontostilbe hasemani, Pimelodella serrata, Prionobrama filigera, Sturisoma nigrirostrum,* and *Thoracocharax stellatus* among others. Station with medium diversity.

Manuripi/Lower Tahuamanu Sub-Basins (Stations P02-28 to P02-46)

Field Station 96-P-02-28
Locality: Río Manuripi above camp to the south ca 9 km.
Latitude and longitude unavailable.
15/Sep/1996
White turbid water river (75 m wide). The shore and bottom are sand/muddy. No aquatic plants. Gallery forest in margins. A total of 175 specimens were collected.
The species list includes:
Characiformes = 14
Siluriformes = 10
Gymnotiformes = 4
Perciformes = 1
Species total = 29
The most abundant species are *Moenkhausia colletti* (51 = 29.1 %), *Moenkhausia lepidura* (37 = 21.1 %), *Hemigrammus* sp. (16 = 9.1 %), *Ctenobrycon spilurus* (11 = 6.2 %), and *Ochmacanthus alternus* (10 =5.7 %). Other species include *Astyanax abramis, Carnegiella strigata, Corydoras loretoensis, Eigenmannia humboldtii, E. virescens, Hypoptopoma joberti, Knodus caquetae, Parotocinclus* sp., *Poptella compressa, Rineloricaria lanceolata, Sternopygus macrurus,* and *Stethaprion crenatum* among others. Station with medium diversity.

Field Station 96-P-02-29
Locality: Río Manuripi at beach 5.78 km from camp, 23 km. from Puerto Rico.
11° 11' 13" S, 67° 33' 20" W
15/Sep/1996
White turbid water river (75 m wide). The shore and bottom are muddy with some aquatic plants and grasses. A total of 216 specimens were collected.
The species list includes:
Characiformes = 13
Siluriformes = 11
Gymnotiformes = 5
Perciformes = 2
Species total = 31
The most abundant species are *Pimelodella gracilis* (46 = 21.3 %), *Hemigrammus* sp. (30 = 13.9 %), *Moenkhausia lepidura* (24 = 11.1 %), *Moenkhausia colletti* (23 = 10.6 %), and *Ctenobrycon spilurus* (11 = 5.1 %). Other species include *Adontosternarchus clarkae, Apistogramma* sp., *Carnegiella myersi, Eigenmannia trilineata, E. humboldtii, E. macrops, Gasteropelecus sternicla, Hemiodontichthys acipenserinus, Hoplias malabaricus, Parotocinclus* sp., *Sternopygus macrurus,* and *Sturisoma nigrirostrum* among others. Station with medium to high diversity.

Field Station 96-P-02-30

Locality: Río Manuripi from below camp to Puerto Rico.

11° 08' 06" S, 67° 33' 20" W

15/Sep/1996

White turbid water river (75 m wide). The shore and bottom are clean mud. Grasses on shore. A total of 75 specimens were collected.

The species list includes:

Characiformes = 4

Siluriformes = 10

Gymnotiformes = 6

Species total = 20

The most abundant species are *Creagrutus* sp. (16 = 21.3 %), *Opsodoras stubelii* (10 = 13.3 %), *Doras* cf. *carinatus* (9 = 12 %), *Eigenmannia macrops* (8 = 10.6 %), and *Eigenmannia virescens* (6 = 8 %). Other species include *Adontosternarchus clarkae, Apteronotus bonapartii, Hemidoras microstomus, Moenkhausia megalops, Pimelodus altipinnis, Rhabdolichops caviceps, Serrasalmus hollandi,* and *Tympanopleura* sp. among others. Station with low diversity, however several species were new for the expedition. (Trawl)

Field Station 96-P-02-31

Locality: Lagoon off Río Manuripi, 1.99 km from camp up river, 5.3 km from Puerto Rico.

11° 09' 06" S, 67° 33' 43" W

15/Sep/1996

White turbid water flooded lagoon. The shore and bottom are muddy with some submerged terrestrial plants. Abundant aquatic plants. A total of 65 specimens were collected.

The species list includes:

Characiformes = 15

Siluriformes = 6

Perciformes = 1

Species total = 22

The most abundant species are *Pygocentrus nattereri* (10 = 15.3 %), *Mylossoma duriventris* (8 = 12.3 %), *Loricariichthys* sp. (6 = 9.2%), *Potamorhina altamazonica* (4 = 6.1 %), and *Potamorhina laitior* (4 = 6.1 %). Other species include *Astronotus crassipinnis, Curimatella myersi, Cynodon gibbus, Liposarcus disjunctivus, Prochilodus nigricans, Psectrogaster rutiloides, Pseudodoras niger, Serrasalmus marginatus,* and *Serrasalmus rhombeus* among others. Station with medium diversity. (Gillnet)

Field Station 96-P-02-32

Locality: Río Manuripi at beach outside lagoon, 1.93 km upriver from camp, 5.3 km from Puerto Rico.

11° 09' 05" S, 67° 33' 40" W

15/Sep/1996

White turbid water river (75 m wide). The shore and bottom are muddy. Some aquatic plants and grasses in margins. A total of 95 specimens were collected.

The species list includes:

Characiformes = 20

Siluriformes = 17

Gymnotiformes = 4

Perciformes = 2

Species total =43

The most abundant species are *Eigenmannia macrops* (98 = 22.6 %), *Pimelodella itapicuruensis* (68 = 15.7 %), *Ctenobrycon spilurus* (58 = 13.4 %), *Cyphocharax spiluropsis* (46 = 10.6 %), and *Pimelodella gracilis* (39 = 9.0 %). Other species include *Auchenipterichthys thoracatus, Ageneiosus caucanus, Apistogramma linkei, Cheirocerus eques, Curimatella dorsalis, Cyphocharax plumbeus, Distocyclus conirostris, Eigenmannia virescens, E. macrops, Hemidoras microstomus, Hemisorubim platyrhynchos, Mesonauta festivus, Ochmacanthus alternus, Pimelodus blochii, Prionobrama filigera, Trachydoras paraguayensis,* and *Triportheus angulatus* among others. Station with high diversity.

Field Station 96-P-02-33
Locality: Río Manuripi at beach 6.36 km upriver from camp, 9.78 km from Puerto Rico.
11° 11' 30" S, 67° 33' 45" W
16/Sep/1996
White turbid water river. The shore and bottom are sandy/muddy. No aquatic plants, but the beach across the river had some grasses. A total of 176 specimens were collected.
The species list includes:
Characiformes = 9
Siluriformes = 9
Gymnotiformes = 3
Species total = 21
The most abundant species are *Moenkhausia lepidura* (55 = 31.3 %), *Moenkhausia colletti* (26 = 14.8 %), *Entomocorus benjamini* (25 = 14.2 %), *Ctenobrycon spilurus* (14 = 7.9 %), and *Hypoptopoma joberti* (10 = 5.7 %). Other species include *Cochliodon cochliodon, Corydoras acutus, Eigenmannia macrops, Knodus caquetae, Rineloricaria lanceolata,* and *Sternopygus macrurus* among others. Station with low diversity.

Field Station 96-P-02-34
Locality: Río Manuripi at beach on E side of river 5.05 km upriver from camp.
11° 10' 49" S, 67° 33' 30" W
16/Sep/1996
White turbid water river (75 m wide). The shore and bottom are muddy with some terrestrial plants. Some aquatic plants. A total of 551 specimens were collected.
The species list includes:
Characiformes = 24
Siluriformes = 13
Gymnotiformes = 1
Perciformes = 4
Species total = 43
The most abundant species are *Moenkhausia colletti* (106 = 19.2 %), *Pimelodella gracilis* (104 = 18.8 %), *Corydoras loretoensis* (88 = 15.9 %), *Apistogramma* sp. (44 = 8.0 %), and *Knodus caquetae* (41 = 7.4 %). Other species include *Anchoviella carrikeri, Brochis splendens, Bunocephalus* sp., *Corydoras acutus, Ctenobrycon spilurus, Eigenmannia virescens, Gasteropelecus sternicla, Hemiodontichthys acipenserinus, Imparfinis stictonotus, Mesonauta festivus, Piabucus melanostomus,* and *Prionobrama filigera* among others. Station with high diversity.

Field Station 96-P-02-35
Locality: Río Manuripi close to Puerto Rico.
11° 08' 06" S, 67° 33' 20" W
16/Sep/1996
White turbid water river (75 m wide). The shore and bottom are muddy with some submerged logs. No aquatic plants. Disturbed forest. A total of 103 specimens were collected.
The species list includes:
Characiformes = 5
Siluriformes = 10
Gymnotiformes = 2
Perciformes = 1
Species total = 18
The most abundant species are *Creagrutus* sp. (46 = 44.6 %), *Eigenmannia macrops* (11 = 10.6 %), *Moenkhausia megalops* (11 = 10.6 %), *Pimelodus blochii* (8 = 7.7 %), and *Megalonema* sp. nov. (8 = 7.7 %). Other species include *Corydoras acutus, Crenicichla heckeli, Eigenmannia virescens, Hemidoras microstomus, Opsodoras humeralis, Pimelodella gracilis,* and *Serrasalmus hollandi* among others. Station with low diversity.

Field Station 96-P-02-36

Locality: Río Tahuamanu near mouth into Río Manuripi, at sandy beach and backwater, 1 km above Puerto Rico.
11° 06' 43" S, 67° 33' 46" W
17/Sep/1996
White turbid water and blackwater mixed. The shore and bottom are muddy. Some aquatic plants in backwater. A total of 220 specimens were collected.
The species list includes:
Characiformes = 20
Siluriformes = 13
Gymnotiformes = 1
Perciformes = 4
Species total = 38
The most abundant species are *Tympanopleura* sp. (35 = 15.9 %), *Prionobrama filigera* (34 = 15.4 %), *Aphanotorulus frankei* (17 = 7.7 %), *Poptella compressa* (12 = 5.5 %), *Pimelodella gracilis* (11 = 5.0 %), and *Engraulisoma taeniatum* (11 = 5.0 %). Other species include *Astyanax abramis, Corydoras loretoensis, Crenicichla heckeli, Doras* cf. *carinatus, Galeocharax gulo, Knodus victoriae, Paragoniates alburnus, Prochilodus nigricans, Serrasalmus hollandi,* and *Thoracocharax stellatus* among others. Station with medium to high diversity. Notice the high density of *Tympanopleura* sp.

Field Station 96-P-02-37

Locality: Lagoon and backwater off Río Manuripi, 1.7 km above Puerto Rico.
11° 06' 57" S, 67° 32' 54" W
17/Sep/1996
White turbid water flooded lake and swamp. The shore and bottom are muddy. Abundant aquatic plants and grasses. A total of 620 specimens were collected.
The species list includes:
Characiformes = 21
Siluriformes = 16
Gymnotiformes = 2
Perciformes = 4
Species total = 43
The most abundant species are *Parotocinclus* sp. (117 = 18.8 %), *Corydoras loretoensis* (97 = 15.6 %), *Cyphocharax spiluropsis* (80 = 12.9 %), *Moenkhausia dichroura* (43 = 6.9 %), *Carnegiella myersi* (42 = 6.7%), and *Ctenobrycon spilurus* (39 = 6.2 %). Other species include *Abramites hypselonotus, Amblydoras hancockii, Anadoras grypus, Aphanotorulus frankei, Bunocephalus coracoideus, Cochliodon cochliodon, Eigenmannia trilineata, Hemigrammus ocellifer, Iguanodectes spilurus, Moenkhausia sanctaefilomenae, Pimelodella gracilis, Poptella compressa, Rineloricaria* sp., and *Serrasalmus hollandi* among others. Station with high diversity.

Field Station 96-P-02-38

Locality: Lagoon and backwater off Río Manuripi, 2.63 km above Puerto Rico.
11° 07' 39" S, 67° 33' 30" W
17/Sep/1996
White turbid water lagoon. The shore and bottom are muddy. Some aquatic plants. A total of 232 specimens were collected.
The species list includes:
Characiformes = 9
Siluriformes = 7
Gymnotiformes = 2
Perciformes = 3
Species total = 21
The most abundant species are *Corydoras loretoensis* (100 = 43.1 %), *Apistogramma linkei* (47 = 20.2 %), *Moenkhausia colletti* (20 = 8.6 %), *Amblydoras hancockii* (9 = 3.8 %), *Eigenmannia virescens* (9 = 3.8 %), and *Ctenobrycon spilurus* (9 = 3.8 %). Other species include *Carnegiella myersi, Corydoras acutus, Eigenmannia humboldtii, Mesonauta festivus, Moenkhausia dichroura, Pimelodella gracilis,* and *Triportheus angulatus* among others. Station with low diversity.

Field Station 96-P-02-39
Locality: Río Tahuamanu near mouth into Río Manuripi, at sandy beach and backwater, 1 km above Puerto Rico.
11° 08' 35" S, 67° 33' 23" W
18/Sep/1996
White turbid water and blackwater mixed. The shore and bottom are muddy. Some aquatic plants in backwater. A total of 389 specimens were collected.
The species list includes:
Characiformes = 16
Siluriformes = 14
Gymnotiformes = 4
Perciformes = 2
Species total = 34
The most abundant species are *Corydoras loretoensis* (116 = 29.8 %), *Moenkhausia lepidura* (67 = 17.2 %), *Pimelodella gracilis* (51 = 13.1 %), *Moenkhausia colletti* (40 = 10.2 %), *Knodus victoriae* (15 = 3.8 %), and *Eigenmannia virescens* (15 = 3.8 %). Other species include *Corydoras acutus, Crenicichla heckeli, Ctenobrycon spilurus, Entomocorus benjamini, Gasteropelecus sternicla, Pimelodella cristata, Serrasalmus hollandi,* and *Vandellia cirrhosa* among others. Station with medium to high diversity. Notice the high density of *Corydoras* that are popular in the aquarium trade.

Field Station 96-P-02-40
Locality: Small cocha on east side of Río Manuripi, 1.5 km above camp, 4.95 km from Puerto Rico.
11° 08' 54" S, 67° 33' 32" W
18/Sep/1996
Brackish water lagoon. The shore and bottom are muddy. Some aquatic plants and grasses. A total of 505 specimens were collected.
The species list includes:
Characiformes = 16
Siluriformes = 11
Gymnotiformes = 1
Perciformes = 6
Species total = 34
The most abundant species are *Corydoras loretoensis* (224 = 44.3 %), *Brachyrhamdia marthae* (47= 9.3 %), *Amblydoras hancockii* (30 = 5.9 %), *Apistogramma linkei* (29 = 5.7 %), and *Ctenobrycon spilurus* (23= 4.5 %). Other species include *Cheirodon piaba, Crenicara unctulata, Hemigrammus ocellifer, Hoplosternum thoracatus, Moenkhausia colletti, Nannostomus trifasciatus, Pimelodella boliviana, Pyrrhulina vittata,* and *Rineloricaria lanceolata* among others. Station with medium to high diversity. Notice the high density of species important to the aquarium trade.

Field Station 96-P-02-41
Locality: Río Manuripi, at small lagoon 5.27 km above Puerto Rico.
11° 09' 03" S, 67° 33' 40" W
18/Sep/1996
White turbid water. The shore and bottom are muddy. Some aquatic plants and grasses in margins. A total of 628 specimens were collected.
The species list includes:
Characiformes =14
Siluriformes = 12
Gymnotiformes = 3
Perciformes = 3
Species total = 32
The most abundant species are *Corydoras loretoensis* (310 = 49.3 %), *Pimelodella itapicuruensis* (96 = 15.3 %), *Pimelodella gracilis* (49 = 7.8 %), *Cyphocharax spiluropsis* (32 = 5.1 %), and *Gasteropelecus sternicla* (17 = 2.7 %). Other species include *Apistogramma linkei, Cheirodon piaba, Corydoras acutus, Eigenmannia virescens, Hoplias malabaricus, Mesonauta festivus, Ochmacanthus alternus, Pimelodella cristata, Rineloricaria lanceolata,* and *Sternopygus macrurus* among others. Station with medium to high diversity. Notice the high density of *Corydoras loretoensis.*

Field Station 96-P-02-42
Locality: Río Manuripi from the camp 3.47 km above Puerto Rico.
11° 08' 06" S, 67° 33' 20" W
18/Sep/1996
White turbid water river (75 m wide). The shore and bottom are muddy. Some aquatic plants and grasses in margins. A total of 12 specimens were collected.
The species list includes:
Siluriformes = 2
Species total = 2
The two species are *Pimelodus blochii* (10 = 83.3 %) and *Opsodoras stubelii* (2 = 16.7 %). (Trawl)

Field Station 96-P-02-43
Locality: Lagoon off Río Manuripi, 0.81 km above Puerto Rico.
11° 06' 39" S, 67° 33' 23" W
19/Sep/1996
White turbid water lagoon. The shore and bottom are muddy. Some aquatic plants and grasses in margins. A total of 1083 specimens were collected.
The species list includes:
Characiformes = 17
Siluriformes = 15
Perciformes = 9
Species total = 41
The most abundant species are *Ctenobrycon spilurus* (223 = 20.5 %), *Amblydoras hancockii* (194 = 17.9 %), *Corydoras loretoensis* (143 = 13.2%), *Hemigrammus lunatus* (105 = 9.6 %), *Hemigrammus unilineatus* (92 = 8.5 %), and *Apistogramma* sp. A (89 = 8.2 %). Other species include *Ancistrus* sp., *Astrodoras asterifrons, Astyanax abramis, Brachyrhamdia marthae, Chaetobranchiopsis orbicularis, Corydoras napoensis, Dianema longibarbis, Cichlasoma severum, Liposarcus disjunctivus, Mesonauta festivus, Moenkhausia colletti, M. dichroura, Pimelodella cristata, P. gracilis, Pimelodus pantherinus, P. blochii, Rineloricaria lanceolata,* and *Satanoperca acuticeps* among others. Station with high diversity. Notice the high diversity of species important in the aquarium trade.

Field Station 96-P-02-44
Locality: Lagoon behind island of camp on NE side, 3.47 km upriver from Puerto Rico.
11° 08' 06" S, 67° 33' 20" W
20/Sep/1996
Whitewater flooded lagoon. The shore and bottom are muddy with logs and leaves. Some aquatic plants and grasses. A total of 421 specimens were collected.
The species list includes:
Characiformes = 10
Siluriformes = 8
Perciformes = 4
Other = 1
Species total = 23
The most abundant species are *Hemigrammus unilineatus* (250 = 59.4 %), *Apistogramma* sp. A (38 = 9.0 %), *Corydoras loretoensis* (27 = 6.4 %), *Pyrrhulina vittata* (25 = 5.9 %), and *Ctenobrycon spilurus* (16 = 3.8 %). Other species include *Aequidens* sp., *Amblydoras hancockii, Cyphocharax spiluropsis, Dianema longibarbis, Hemigrammus ocellifer, Iguanodectes spilurus, Rineloricaria lanceolata,* and *Tridentopsis pearsoni* among others. Station with low diversity. Notice the high density of *Hemigrammus unilineatus.*

Field Station 96-P-02-45
Locality: Río Manuripi, in front of camp 3.47 km above Puerto Rico.
11° 08' 06" S, 67° 33' 20" W
20/Sep/1996
White turbid water river. The shore and bottom are muddy. Some aquatic plants and grasses in margins. A total of 349 specimens were collected.
The species list includes:
Characiformes = 15
Siluriformes = 7
Gymnotiformes = 1
Species total = 23
The most abundant species are *Moenkhausia colletti* (74 = 21.2 %), *Corydoras loretoensis* (69 = 19.8 %), *Moenkhausia lepidura* (67 = 19.2 %), *Ctenobrycon spilurus* (44 = 12.6 %), *Pimelodella gracilis* (32 = 9.1 %), and *Prionobrama filigera* (14 = 4.0 %). Other species include *Carnegiella myersi, Eigenmannia virescens, Moenkhausia dichroura, Pimelodella cristata, Poptella compressa,* and *Stethaprion crenatum* among others. Station with medium to low diversity.

Field Station 96-P-02-46
Locality: Río Manuripi, lagoon SW side of island 3.47 km upriver from Puerto Rico.
11° 08' 06" S, 67° 33' 20" W
20/Sep/1996
White turbid water flooded lagoon. The shore and bottom are muddy. Some aquatic plants and grasses in margins. A total of 54 specimens were collected.
The species list includes:
Characiformes = 5
Siluriformes = 4
Perciformes = 2
Species total = 11
The most abundant species are *Cyphocharax spiluropsis* (15 = 27.7 %), *Amblydoras hancockii* (13 = 24.1 %), *Astrodoras asterifrons* (7 = 13 %), *Parotocinclus* sp. (7 = 13%), and *Curimatella dorsalis* (4 = 7.4 %). Other species include *Auchenipterichthys thoracatus, Apistogramma linkei, Hoplias malabaricus, Mesonauta festivus,* and *Moenkhausia colletti.*
Station with low diversity.